Wiring 12 Volts

for

Ample Power

(Revised 1995)

David Smead and Ruth Ishihara

Illustrations by Mitch Ishihara

RIDES Publishing Company
Seattle, Washington

Published by:

RIDES Publishing Company
2442 NW Market Street #43
Seattle, Washington 98107 U.S.A.

First printing 1990
Second printing 1991
Third Printing 1992
Fourth Printing 1993
Fifth Printing (Revised) 1995

Library of Congress Catalog Card Number: 89–92666
ISBN 0–945415–03–6

About the Authors and Illustrator

David Smead and Ruth Ishihara are the authors of the marine best seller *Living on 12 Volts with Ample Power.*

David Smead is an engineer, computer programmer and the builder of a 50′ ketch. His engineering background began in 1963 as an analog engineer designing digital voltmeters. In the late 1960's and early 1970's, he supervised engineers and programmers building automatic test equipment and process control systems.

From 1973 to the present, David has been self-employed, obtaining contracts for hardware and software design from major corporations. He is the electronic engineer and programmer for Ample Power products, and a leading electrical system designer for marine, RV and remote home applications.

David and Ruth spent two years living on their 12 Volt system while cruising in Mexico and Hawaii.

Ruth Ishihara is a research and word processing specialist. She was employed thirteen years for the state of Washington in the legislative, financial and transportation branches preparing documents for various studies, research data and bill proposals. Currently, she is President of Rides Publishing Company.

Mitch Ishihara is studying electrical engineering and computer programming at the University of Washington. He is a PC and Windows[1] expert and used Corel[2] Draw for the illustrations.

[1] Windows is a trademark of Microsoft, Inc.

[2] Corel Corporation

Acknowledgement

The information in this book has been derived from personal experience, and the references listed in the bibliography. To cite each instance of a factual source would have resulted in a book of citations, without the necessary coherence to amplify details pertinent to small energy systems. All of the references contain valuable information. For those with further interest in the subjects herein, we urge you to consult the bibliography.

We thank Michael McGee for his advice, examples of style, and his input on several chapters.

Preface to the Revised (1995) Edition

A lot of changes have occured in the five years that have passed since the original book was published. Electrical systems have evolved so that more and more people have discovered the benefits of thoughtful design instead of the hodge–podge of systems which fell out of the automobile.

While considerable progress has occurred, electrical systems are still catching up to the ideas we first presented in *Living on 12 Volts with Ample Power* and the first edition of this book. More manufacturers are now providing multi–step chargers and regulators, and a few are even providing temperature compensation.

Much of the contents of this book remain the same. Information on batteries has been expanded, although the prime reference on batteries is still *Living on 12 Volts with Ample Power.* A new chapter on system design has been added which will help the reader integrate some of the other detailed information into a system plan. The chapter on components has been expanded and some parts clarified. Changes, additions and clarifications have been made throughout the other parts of the book.

About Publishing

In 1986 when we first started to finalize the details of *Living on 12 Volts with Ample Power*, desktop publishing was an infant. We tried a well known software package of the era and, in short order, were thoroughly disgusted with its performance. At that time we read about a publishing program called TEX that originated at the University of Stanford. It was authored by a reknowned computer scientist, Donald Knuth, and was available free for those running the Unix operating system. We decided to try a DOS version for a quite modest price.

Within a week of getting the program installed on a very early PC with a whopping 10 Mb hard drive, we had the first book formated in the desired style. Wow!

In the end, however, it wasn't all that simple. First we had to write our own program to sort the index file into a meaningful format. Then we learned about pasting and stripping figures into a final copy of the text. Every *run* of the program took about an hour and a half!

Nine years later we can run the book through the program in under two minutes, using a considerably faster computer, and an improved version of the publishing program. Figures and photographs are no longer glued to a page of text and then photographed to make printing plates. Photographs are scanned into the computer for direct output with the text. Drawings are likewise made using computer drafting tools and printed along with the text. Cameras aren't needed now to make plates ...output from the computer can now be sent to *imagesetters* which print high resolution film directly.

TEX and its *packages* that implement certain publishing functions are maintained by a few good people who know it well ... most of them in the university system around the world. It is available free from the Internet in the CTAN archive at Sam Houston State University. There is a TEX Users Group at POB 869, Santa Barbara, CA 93102–0869. We use LATEX, a package by Leslie Lamport that runs on top of TEX.

While not a publishing program for everyone, TEX and LATEX have been of immense assistance to us and many others who have used it to publish books, catalogs, letters and other documents. We think that TEX and its derivatives are a first class example of good ... by unselfish dedication a tool was designed and passed on to enrich the lives of others. Enhancement and maintenance of the tool is a cooperative, ongoing effort all over the world.

This book has gone from computer to film without any manual steps in between. While the intermediate work of writing the text and drawing figures or scanning photographs hasn't been any easier, the fact that it can be viewed exactly as it will appear in print with all the figures and photos in place has made it easier to polish the presentation. What ever imperfections remain are the fault of the authors ... we haven't lacked for the right tool.

Contents

List of Tables

List of Figures

Chapter 1

Introduction

1.1 General Information

If you are building a boat, RV, or remote home from scratch, then this book will get you started on the electrical system in a competent style.

If you already own a battery powered system, chances are high that sooner or later you will decide to upgrade it. Having seen the electrical workmanship in many different boats and RVs, and after observing the quality of batteries, charging equipment and the like, which is typically installed by the manufacturers, we are certain that the systems will not perform to your satisfaction.

Sometimes a few small changes can turn a non–functional system into a reasonable power supply. As often as not, major replacement of components with complete rewiring is necessary. Whether you do the system upgrade yourself, or hire someone to do the work for you, this book explains how to plan and implement the task, what materials are suitable, what tools are required, how to use the tools, and where to buy tools and material.

While we've made an attempt to keep this book complete, with no other material necessary, we do refer to our first book, *Living on 12 Volts with Ample Power*. It remains the definitive reference for batteries, alternators, chargers and refrigeration. It also provides great details about the design of a system. The chapters in this book on batteries and charge sources have some redundant information, but do not provide in–depth coverage.

1.2 About the Chapters

The order of presentation is often fixed by the subject matter. This is only partially the case for the subjects covered. We decided to put a historical chapter, Chapter 2, before the subject of electricity was presented. All too often, we are given the full details of a subject as if it were discovered all at once and wrapped up with a few equations. The history of science and technology is a delightful subject. Once you understand the chronological order of discoveries, their dependencies, and the crude apparatus used to test observations, what seems like an unfathomable subject becomes a little simpler.

1.2.1 Chapter 3, 4 and 5

The history of electrics, is followed by basic DC electrical theory, and then DC magnetics. These subjects have been well understood for over a century and need not hold any mystery for a person of the 1990s. DC theory is followed by a chapter on AC. While the battery powered system is primarily a DC system, AC is still used.

1.2.2 Chapter 6

Chapter 6 is about loads ...lights, computers, fans, inverters, etc. The system exists to power loads, and until the loads are defined, no accurate choices about batteries or chargers can be made.

1.2.3 Chapter 7

With loads understood, the next subject of interest is batteries. There must be enough battery power to supply load current for as long as necessary between recharges. Batteries are not as simple as automotive use appears to make them. Deep cycle batteries are expensive, and will only provide good service if maintained correctly.

1.2.4 Chapter 8

Batteries must be recharged after use, so the next chapter covers charge sources ...solar panels, alternators, AC chargers, and wind/water generators. All of these devices need to be regulated to match battery characteristics.

1.2.5 Chapter 9

A independent power system can not be properly managed without instrumentation. A whole chapter is devoted to the subject. The measurement of voltage is critical, but so is the measurement of Amps. A modern system will also include instrumentation of Amp hours consumed, and Amp hours remaining in the batteries.

1.2.6 Chapter 10

With a background in batteries, loads and charge sources, it's time to consider the steps of a system design. Determining loads and battery capacity is described.

1.2.7 Chapter 11

There's more to the energy system than loads, batteries, sources, and instruments. A chapter covers electrical system components such as switches, relays, isolators, regulators, and fuses.

1.2.8 Chapter 12

A separate chapter describes wires, terminals, and wiring aids, explaining what is and isn't appropriate.

1.2.9 Chapter 13

Next, comes the subject of tools. No job can be accomplished without tools. While pros with years of experience may perform acceptable work with a minimal toolset, the right tool is essential for the amateur.

1.2.10 Chapter 14

After the subject of tools, a chapter of schematics is presented. Many of the schematics may be used as presented. Others may need some modification for your particular system.

1.2.11 Chapter 15

With an understanding of the system, the instruments required to operate it, the components needed, the wire and terminals necessary, and the tools to accomplish the installation, you are ready for the chapter about wiring practices. Wire routing is never an easy task, but will be a little simpler once you have digested the chapter on wiring practices.

1.2.12 Chapter 16

A concluding chapter covers the subject of testing. Unless you do a perfect wiring job, some amount of testing will be necessary after wiring. Testing is always required when trouble strikes.

Chapter 2

Pursuit of the Elusive Electron

Thales[1] is on record about 600 B.C. for posing the question *what is the universe made of?*. He concluded that the most basic component was water, since life relied so much on it. Thales is thought to be the first to study magnetism. If lodestones were made of water, Thales couldn't explain how, but his failings are not as important as his question. Humans have spent countless fortunes building apparatus to answer Thale's question, and only recently is there strong evidence for three fundamental *families* of basic constituents.

That the universe could be comprised of basic elements was an important concept. Why does a magnet attract iron at a distance? What kind of invisible forces are at play?

About 500 B.C., Democritus[2] proposed that all matter was made from minute, indivisible particles. The Greek word for indivisible is *atom.*

William Gilbert[3] extended knowledge of magnetism. Units of magnetic force are called *gilberts* in his honor. Gilbert also experimented with static electricity. Early Greeks had known that rubbing amber gave it the power to attract light weight objects, as opposed to iron which was attracted by magnets. Gilbert discovered that rock crystals and certain

[1] Greek philospher, 624 B.C. – 546 B.C.
[2] Greek philospher, 470 B.C. – 380 B.C.
[3] English physician and physicist, 1540 – 1603.

gems acted like amber. The Greek word for amber is *elektron*, so Gilbert coined the word *electrics* for amber–like material.

Otto von Guericke[4] mechanized the generation of static electricity. His sulphur globe was driven by a crank, and when stroked, the globe would take on a charge. Later it could be discharged, producing sparks.

In 1729, Stephen Gray[5] discovered that electric charges could be conducted, and shortly thereafter the words *conductor* and *insulator* were added to the English language by John Desaguliers[6] He chose insulator from the Latin word *island.*

Peter van Musschenbroek[7] invented a device in 1746, called a Leyden jar, that could store static electricity. The jar was a metal container suspended by silk cords. Water in the container was electrically coupled externally via a brass wire through a cork. An assistant to Musschenboek discovered how much energy could be stored when he accidently discharged the jar through his body. He lived, but the event was nevertheless newsworthy.

By now, a lot was known about electricity. It could be generated by friction, it could be conducted, and it could be stored. As we all know, static electricity can give quite a jolt, but only for an instant. Static electricity could be produced by a sulphur globe, and Leyden jars could store electricity, perhaps enough to be lethal, but they too quickly discharged. Was there a source of electricity that could sustain a discharge for long periods?

Benjamin Franklin[8] used a Leyden jar in his famous kite experiment. Franklin, as the fifteenth child of seventeen, had only two years of formal education. In 1752, he flew a kite during an electrical storm which had a wire leading down to Franklin. By placing a key near the wire, he could draw sparks[9]. Later he charged a Leyden jar, proving quite simply that the atmosphere is charged. But with what kind of particles?

In 1766, Franklin was in London, trying to resolve the taxation issue

[4]German physicist, 1602 – 1686.

[5]English electrical experimenter, 1696 – 1736.

[6]French–English physicist, 1683 – 1744.

[7]Dutch physicist, 1692 –1761.

[8]U.S. statesman and scientist, 1706 – 1790.

[9]The next two persons attempting corroboration of Franklin's experiment were electrocuted ... you probably shouldn't try to collect electrons from the sky.

for the colonies. There he met a man by the name of Joseph Priestley[10]. Priestley was impressed by Franklin and took up a career in science. He discoverd that carbon was a conductor, and later suggested that electricity would play a large part in the field of chemistry. Priestley took up chemistry and made many important discoveries about gasses[11].

Charles Coulomb[12] showed in 1785 that the force of electrical attraction or replusion between charged spheres is proportional to the product of the charge on each sphere, and inversely proportional to the square of the distance between them. Isaac Newton[13] had shown that gravity works on an inverse square law. Was electricity the same as gravity? Coulomb used his invention, a torsion balance, to measure attraction or repulsion, and therefore to establish the amount of charge on a sphere. The unit of electrical charge is the Coulomb, in his honor.

The synthesis between electricity and chemistry, that Priestley had predicted, began in 1790 when Luigi Galvani[14] discovered that two dissimilar metals in blood could generate enough electricity to cause muscle twitches in frogs. Though Galvani didn't understand the phenomena, we still use the phrase *galvanic action* to refer to electrical currents between metals in an electrolyte.

Alessandro Volta[15] followed the work of his friend, Galvani. In 1794, he discovered that animal blood wasn't necessary for galvanic action. By 1800, Volta was making *batteries.* The word battery then, and now, means a group of similar things used together, such as a battery of guns, or a battery of scholastic tests. An electric battery is most often a group of similar cells. By providing the world with a reliable source of electricity, Volta threw open the gates to electrical research. Could it be long before people knew everything there is to know about electricity?

John Dalton[16] advanced the atomic theories of ancient Democritus. Whereas Democritus expounded atomic theory with no supporting evidence, Dalton observed that compounds of two elements always appeared in whole number ratios such as 3 to 8, or 6 to 1, and never appeared in

---

[10] English chemist, 1733 – 1804.

[11] Priestley was the first to dissolve carbon dioxide in water ... tonic water.

[12] French physicist, 1736 –1806.

[13] English scientist and mathematician, 1642 – 1727.

[14] Italian anatomist, 1737 – 1798.

[15] Italian physicist, 1745 – 1827.

[16] English chemist, 1776 – 1844.

fractional ratios such as 3.9 to 1. By working out the proportions of weights of various compunds, Dalton was able to construct the first table of atomic weights. His theory of 1803, that all atoms in a given element were identical, but atoms in different elements were different, was readily accepted.

Pieces to the puzzle of matter were indeed coming together in the chemical arena. But if all of matter consisted of indivisible atoms, where did electricity fit? Was there an electric atom, perhaps the basic particle in a magnet?

Hans Oersted[17] discovered in 1819 that electric current deflected the magnetic needle of a compass. The connection between electricity and magnetism was clearly established, and furthermore, electric current could now be quantized. Whereas Coulomb had provided the means to measure the amount of electric charge stored on a sphere, it was now possible to measure the amount of charge flowing from one point to another. The amount of compass needle deflection indicated the amount of current flow. A week after Oersted's discoveries were reported, André Ampère[18] developed a *lines of force* theory for magnetism and went on to quantify electric flow. The unit of current flow became the Ampere.

Ampère went on to show that magnetic force could be attained without magnets. Two parallel wires carrying current attracted each other when they carry current in the same direction, and repulsed each other when the current flowed in opposite directions. He showed that wires wound in a spiral, or helix behaved like a magnet.

William Sturgeon[19] developed Ampère's idea of a helical magnet further. In 1823, Sturgeon could lift nine pounds with his electromagnet ... wires wound around an iron core. By 1831, Joseph Henry[20] had an electromagnet that could lift 750 pounds.

Conductors and insulators were known. In 1827, Georg Ohm[21] described *resistance* to electrical flow. Ohm showed that current flow was proportional to the potential difference across a conductor, and inversely proportional to the resistance of the conductor. In other words, the higher

[17] Danish physicist, 1777 – 1851.
[18] French mathematician and physicist, 1775 – 1836.
[19] English physicist, 1783 – 1850.
[20] American physicist, 1797 – 1878.
[21] German physicist, 1789 – 1854.

the amount of charge, the greater the current, and the higher the resistance, the less the current. The unit of electrical resistance is called an Ohm in his honor.

In the meantime, about 1821, Michael Faraday[22] invented the first electric motor. In the 1830's, Faraday performed the quantitative analysis of electrochemical reactions that take place in a battery. Faraday originated the terms, *electrolyte, electrode, anode,* and *cathode.*

Faraday had discovered that iron filings would align themselves in a peculiar fashion between the poles of a magnet. This let him *see* magnetic lines of force. The lines of force are stationary for a static magnet, but for an electric current building up in a wire, or decaying, the lines of force would have to grow or shrink. He went on to invent the transformer, a device that couples electricity from one coil of wires to another when the current is varied in one coil. The changing lines of force from the primary wires *induced* a current in the secondary wires.

Between 1864 and 1873, James Clerk Maxwell[23] provided a mathematical foundation for Faraday's lines of forces. With a few simple equations, he was able to express all of electric and magnetic phenomena. He calculated the speed at which a magnetic field is radiated from its source, and found it to be the speed of light! This suggested to Maxwell that visible light is just part of a spectrum of electromagnetic forces. It may interest you to know that Maxwell's equations were not modified by Einstein's[24] work which threw out all the rest of classical physics.

Despite knowing how to generate electricity, how to store it, and how to make it perform motorized work, the question still remained, what was the basic electric *thing*? Was it a particle, or some other mysterious force?

Faraday had noticed that electric current forced through a vacuum produced a fluorescent effect, but vacuums of the day were not good, so research in that area was limited.

In 1855, Heinrich Geissler[25], a talented glassblower, came up with a method of evacuating glass tubes. With a better vacuum, electric research

[22]English physicist and chemist, 1791 – 1867.

[23]Scottish mathematician and physicist, 1831 – 1879.

[24]Albert Einstein, German–Swiss–American physicist, 1879–1955

[25]German inventor, 1814 – 1879.

could advance on another front. In 1876, Eugen Goldstein[26] named the luminescent rays leaving the cathode as *cathode rays.* Today the cathode ray tube is the basis of television and computer terminals.

Goldstein also discovered rays going in the opposite direction from the cathode rays. These rays he named channel rays. Jean Perrin[27] determined in 1895 that cathode rays would negatively charge a cylinder by striking it. Channel rays, on the other hand, were found to be positively charged.

By the 1870's, William Crookes[28] had developed vacuums with pressure only 1/75,000 that of Geissler's tubes. He was able to show that cathode rays could be deflected by a magnet ... almost convincing evidence that the rays were charged particles. The laws of electrochemistry that Faraday had worked out seemed to favor the idea that electricity was not a continuous fluid, but was rather based on a particle that had some specific charge.

In 1891, George Stoney[29] proposed that the electric particle be called an electron. At the time it wasn't clearly established that electricity was a particle.

Joseph John Thomson[30] took up the pursuit of cathode rays in a vacuum. By 1897, Thomson was able to demonstrate that in highly evacuated tubes, cathode rays could be deflected by an electric field, as well as a magnetic field. Electricity is a particle! The name electron stuck. Thomson further showed that the mass of the electron was about 1/1837 that of the lightest atom, hydrogen. The electron is subatomic! Thomson considered the atom to be an essentially featureless sphere, with enough electrons stuck to its surface to balance the positive charge of the sphere.

Other evidence converged about this time to convince even the doubters that the atom was not a motionless sphere, but was made up of moving particles. Wilhelm Roentgen[31] was experimenting with cathode rays in late 1895 when he observed a luminescent paper glowing as it would when cathode rays impinged on it. But, cathode rays could not be hitting the

---

[26] German physicist, 1850 – 1930.
[27] French physicist, 1870 – 1942.
[28] English physicist, 1832 – 1919.
[29] Irish physicist, 1826 – 1911.
[30] English physicist, 1856 – 1940.
[31] German physicist, 1845 – 1923.

paper, because they were blocked by cardboard. What was it? What were these rays that didn't respond to either electric or magnetic fields? It's customary for mathematicians to label the unknown with the letter X. Today we still know the rays as X rays.

Roentgen's X rays aren't germaine in our pursuit of the electron, but indirectly it led to the discovery of radiation from uranium, which was discovered by Antoine Becquerel[32]. Like X rays, the radiation penetrated solids. Becquerel found that at least part of the radiation could be deflected by a magnet. This led him to believe that he was seeing the same electrons that Thomson had discovered. Electrons were being emitted from uranium! This meant that atoms were *divisible*, and must contain electrons.

Ernest Rutherford[33] followed the work of Becquerel. He decided that radiation was a composite of different phenomenon, and discovered positive and negative rays, calling them alpha and beta rays respectively. In 1900, he found that some of the radiation was not affected by a magnet, but was instead an electromagnetic wave. He named that radiation, gamma rays. By 1908, he was firing alpha particles into thin sheets of metal ...gold, 1/50,000 of an inch thick. Most of the alpha particles passed straight through the gold, to register on a photographic plate behind the foil. A few of the alpha particles changed directions in the foil, by more than 90 degrees, in some cases. The deflected particles were hitting something heavy. In 1911, Rutherford announced his theory that an atom is constructed of a very small, but dense nucleus at the center, and the nucleus is surrounded by a cloud of light electrons that don't impede the passage of alpha particles.

Positive, or channel rays had been discovered earlier. They were much like alpha rays. In 1914, Rutherford felt that the simplest of positive rays were those of a fundamental, positively charged particle. He named the particle the proton. Rutherford correctly declared that the nucleus of an atom is comprised of protons. Neutrons were discovered later, and particle discovery is still predicted today by some rather esoteric mathematics.

With Rutherford's discoveries, the electron was finally located, and shown to be but a part of the atom. A cloud of electrons surrounding

[32] French physicist, 1852 – 1908.
[33] British physicist, 1871 – 1937.

a nucleus quickly led to the idea of electrons orbiting the nucleus ... a miniature solar system. Niels Bohr[34] worked first for Thompson and later for Rutherford. He used spectrocopic data to work out a model of the hydrogen atom, which only has a single electron. Earlier, Max Planck[35] had worked out a simple equation that accurately described radiation over the whole spectrum of frequency. His equation relied on the assumption that energy is not infinitely divisible, but instead comes in discrete lumps, which he called *quanta.* Niels Bohr used the quantum theory of Planck in his model of the hydrogen atom. The electron couldn't just orbit at any distance from the nucleus, rather it was constrained to fixed distances. The distance from one orbit to another depended on the quantum of energy that the atom absorbed or released.

Bohr wasn't able to model atoms with more than one electron, but he observed that if there was more than a single electron in orbit, the orbits would be like *shells.* He pointed out that only the electrons in the outer shell determined an atom's chemical properties. You may recall the term, valence, from chemistry classes.

With over 2000 years of electrical history in a few paragraphs, we have obviously omitted much detail, and passed over names of many scientists who put pieces of the puzzle together. We stop our pursuit of the elusive electron here with Bohr, in the year 1913, because the knowledge gained is sufficient to explain the basics of DC and AC electricity. Since 1913, the quantum theory has proven as interesting and accurate as it is counter intuitive. There is still much to be learned about the atom, and the basic forces that bind the universe.

What we have learned is this:

- Electrons are negatively charged constituent particles of atoms.
- Electrons can be separated from atoms by friction.
- Electrons can be stored on charged spheres (or plates).
- The quantity of electrons stored can be measured (Coulomb).
- Electrons can be conducted and isolated (insulated).
- Conductors of electrons can resist the flow.
- A greater charge of electrons can result in greater flow.

[34] Danish physicist, 1885 – 1962.
[35] German physicist, 1858 – 1947.

- The outer most electrons in an atom can be involved in chemical reactions.
- The flow of electrons causes magnetic fields.
- A changing magnetic field can induce current without direct connection.
- Magnetic fields are radiated from their source at the speed of light.
- An atom is mostly void of matter.
- Positively charged particles are called protons.
- Protons are a part of the nucleus.
- Electrons orbit the nucleus at *quantum* distances.
- Electrons weigh about 1/1837 of a hydrogen atom.
- Protons weigh 1836 times an electron ...almost the weight of the hydrogen atom.

The application of what's known is always more difficult than memorization of a few facts. By pursuing the elusive electron, we hope that you have gained an insight into application of electrical phenomena. It can't be that difficult ...it only took 2500 years of the best human minds to catch up with the elusive electron.

Chapter 3

DC Electricity

3.1 Introduction

Understanding **D**irect **C**urrent electricity is much simpler than imagined. While you can't see electrons, the particles that operate electrical machinery, their behavior can be described completely. With modern instruments it is easy and inexpensive to measure the force of electrons and the work that they accomplish. Put aside your mental blocks ...the journey into electrics is fun and rewarding.

3.2 Formal Systems

Before jumping into the study of DC electricity, a note about formal systems is in order. A formal system is one that contains objects, relationships and operations. An object need not be a physical object, it may be just a conceptual entity, such as a square foot. A relationship is simply a description that binds objects together. In a human sense, a relationship may be mother or brother. In other circumstances, a relationship may be bigger, or darker. An operation is an action that can involve the objects. It may be as simple an operation as shaking hands.

What makes a system formal, is this ...operations are constrained to actions that can occur without violating the object or relationships. If we have a couple of rocks that are related by virtue of a common ore content, its extremely unlikely that we can have an operation such that one of them speaks to the other. If rocks start talking, we do not have a formal system.

Stretching the example to such a limit illustrates the point, but to be more realistic, suppose the rocks have been shown to be of exactly the same ore. Would we be suspicious if one melted at 400° F, while the other took 1200° F to melt?

The most formal of formal systems are those of mathematics. Simple arithmetic has objects called numbers. Relationships between numbers are *less than*, *equal*, or *greater*. Operations include addition, subtraction, multiplication, and division. There are even relationships between the operations. For instance, the product of 7 times 8, added to 5 is 61, while the product of 7 times the sum of 8 and 5 is 91. In other words, operations may have a precedence relationship that affects the final outcome. For any given set of objects and operations, the outcome is invariably the same in a formal system.

While there are people that choose to believe humans have a holistic knowledge that defies containment within a formal system, advances of the human race has been through the pursuit of order out of seeming holistic chaos. The discovered order is eventually expressable within the context of a formal system. We note that a formal system can not say anything meaningful about itself, that is, mathematics can not prove that mathematics is complete.

Since Maxwell, electricity has been a formal system. There are objects, and those objects have certain relationships. There is also a specific set of operations which prescribe interaction between the objects. Learning DC electrical theory will be a matter of learning objects, relationships, and operations which are applicable.

3.3 What Electricity Is

Electricity is the movement of electrons through a conductor. Electrons, which are part of atoms, are negatively charged particles which orbit around a positively charged nucleus. In conductors, such as gold, copper, aluminum and other metals, electrons can easily be forced to break orbit and move to a new orbit at an adjacent atom. Each electron leaving an atom is replaced by another in a musical chairs affair. What causes an electron to jump orbits is a difference in electrical potential.

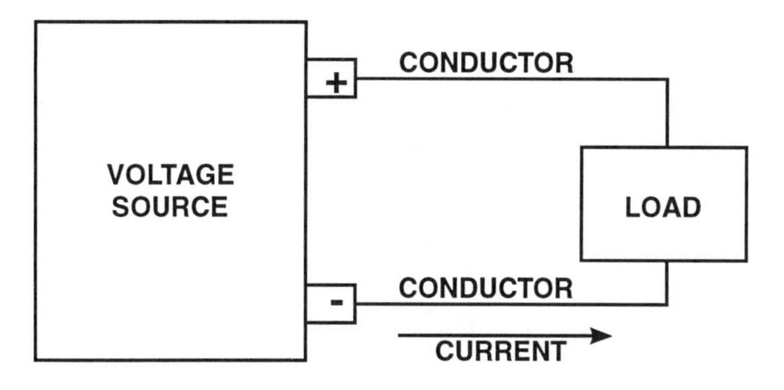

Figure 3.1: A Closed Circuit

Of all there is to know about electricity, the one fact that you must always remember is this; *electrons can only flow in a closed circuit.* Electrons don't run out the end of wires. If they did, the carpet below a household outlet would be piled high with them. An electron that jumps an orbit must be replaced, and it takes a closed circuit to do that.

Figure 3.1 shows what we mean by a closed circuit. The circuit consists of a voltage source, a load, and two conductors which we might call the positive and negative conductors. The direction of current flow is labelled ... it flows from the negative terminal of the voltage source, to the positive terminal. Without the conductors, current could not flow.

Consider a mountain lake. In times of drought, the lake may dry up. Without rain, there isn't a closed circuit, so eventually the water will quit flowing.

It is the movement, or flow of electrons that perform useful work in any electrical device, either by generating heat, or creating a magnetic field. Heat generated by electricity is found in such appliances as ovens, lights and TVs, while magnetic fields are used in motors, horns, relays, and also in TVs.

3.4 Potential Differences

A potential difference causes electrons to jump orbit. Consider two water towers which are on the same level. One of the tanks is full, while the other is empty. If a valve at the bottom of the tanks is opened between them, the water levels will eventually become the same. The force that causes water to flow from the full tank to the empty tank is the difference in pressure.

A potential difference of electrical charges results in the same action. Take two spheres suspended from strings. If one of them is negatively charged with static electricity, while the other is not charged, the spheres will attract each other. On touching, the spheres will become equally charged, and thereafter repel each other. Like charges repel, unlike charges attract, are two sides of a relationship between electric entities.

When one apparatus has more electrons than another apparatus, a potential difference exists between them. Potential difference is measured in Volts, and can easily be measured with a voltmeter.

3.5 Current Flow

As mentioned, electricity is the flow of electrons, which flow because of a potential difference. The flow is called current, not unlike current in a stream of water. The rate of flow is measured in Amperes, or Amps, which is simply a count of how many electrons pass a given point in one second.

A closed circuit is necessary for current flow. Circuits are closed with conductors. Most metals are conductors, although the oxides that form on metal surfaces are not conductors. Material that doesn't conduct is called an insulator. Some common insulators are glass, wood, dry air, rubber, quartz, mica, bakelite, and most plastics.

It is a common misconception that electrons move at the speed of light. The effect of electron movement is felt at the speed of light, but electrons move considerably slower. We can almost say that they *drift* through conductors. When an electron jumps an orbit, another arrives to take its place. By repeating this phenomena, billions of times in a length of wire, it would appear that an electron moved from one end of the wire to the other at the speed of light. You might think of electrons as an on–edge row of dominoes. The effect of one falling is soon felt at the far end, although the dominoes didn't move substantially.

3.6 Voltage Sources

A voltage source is a means of achieving a potential difference. All voltage sources have a positive terminal and negative terminal. By connecting a conductor between the terminals, a potential difference is created from one end of the conductor to the other. The potential difference causes current to flow. Electrons leave the negative terminal of the voltage source, and flow through the conductor to the positive terminal. To sustain a continuous current, the voltage source must have a means of recycling electrons at its positive terminal in order to generate electrons at its negative terminal.

By convention, the positive terminal of a voltage source is colored red, while the negative terminal is colored black. It is customary to use red and black wires for positive and negative, respectively.

There are quite a few fundamental voltage sources. Wires moving through a magnetic field develop a voltage. Dissimilar metals when heated, the thermocouple, produce a small voltage. The sun, striking a solar panel generates voltage. Some crystals, such as those used for hi–fi cartridges produce a voltage from pressure input. Voltage is even produced across metal which is strained by torque.

A battery is a familiar voltage source used to propel electrons through a wide variety of devices. As a voltage source, it has both a positive and a negative terminal. The battery recycles electrons at the positive plate by moving them chemically to the negative plate. A battery is a direct current, or DC voltage source. A direct current voltage source would be better named as a *one way* voltage source, since current always flows through the load in the same way ...from the negative terminal of the voltage source, through the load to the positive terminal of the source.

Sometimes you hear that current flows from positive to negative. This idea dates back to Benjamin Franklin. He considered that electricity was a sort of fluid, and he suggested that an excess of that fluid be called positive, while a deficiency be called negative. The fluid would flow from an excess to a deficiency. When the negatively charged electron was finally isolated, Franklin's suggestion was opposite, since current flows from an excess of negative. No harm is done by believing that current flows from positive to negative ...as long as you remain consistent, you'll get the same result.

Even the best of batteries can not supply a continuous current. Their ability to recycle electrons collected at their positive plate to their negative plate diminishes as current flows. Eventually they discharge to a point where they are no longer capable of supplying sufficient energy to power the load. In this regard, batteries are rated in Amp hours, that is, how many Amps they can supply for a given number of hours. A chapter on batteries gives more details.

3.7 Voltage and Current Relationship

Since voltage causes current to flow, it follows that an increase in voltage will result in an increase of current flow. In the case of a garden hose, an increase in water pressure causes more water to flow, while a decrease in pressure reduces the amount of water.

The relationship of current flow relative to voltage is illustrated by the familiar flashlight. With a new set of batteries, the light is bright. New batteries have a higher voltage than old ones, and that higher voltage causes more current flow. With more current, the filament in the bulb glows hotter and emits more light. As the battery discharges, its voltage declines. Less current flows, and the light gets dimmer, eventually going out when the battery no longer represents a voltage source.

3.8 Resistance

No conductor is perfect, and that's good. Consider water dropping off a water fall. The trip is short, and the landing violent. Any increase in the availability of water at the top shows up nearly instantly at the bottom of the falls. Attempting to regulate the flow of a falls by placing an obstruction half way down would be futile. Contrast the water fall with a twisting rocky stream. Each bend, and each rock imposes a resistance to the flow of water. Streams can even be dammed in order to control the flow. A water fall, on the other hand, is a *short circuit.*

Every conductor has some amount of resistance to current flow. Naturally, some conductors offer more resistance than others. Resistance is absolutely necessary in an electrical circuit. Without resistance, all circuits would be short circuits, and excessive current would flow, causing high temperatures to destroy circuit devices. Sparks can even be produced by short circuits, causing nearby material to ignite.

Resistance acts as a restriction to current flow, not unlike stepping on a garden hose. In the flashlight, the filament in the bulb acts as a resistance. It is the resistance of the filament that generates heat, and therefore light.

Resistance can be thought of as friction. Press you hands firmly on a table top and rub them vigorously to and fro. Soon, your hands will get warm from the friction, or resistance to the motion.

Electrical resistance is measured in units of Ohms.

3.9 Ohm's Law

So far, we have objects including insulators, conductors, resistance, voltage sources and current. Since all conductors contain resistance, we can omit conductors as an object, and instead, consider conductors as a special low resistance form or a resistor. What is the fundamental relationship between voltage, current and resistors?

You may recall that Georg Ohm first expressed this relationship, which has become known as Ohm's law. In words, we can say that Current, (Amps) in a circuit equals Voltage, (Volts) divided by Resistance, (Ohms). To express Ohm's law in conventional symbols, you need to take a great leap of terminology. Current, although measured in Amps, is represented by the letter I. That's because current can *induce* a magnetic compass needle to move. Voltage, measured in Volts, is represented by the letter E, which is short for *electromotive force.* Resistance, thank goodness, is represented by the letter R. Ohm's law is thus stated as:

$$I = E / R$$

Given the resistance in the circuit, and the voltage applied to it, current can be found readily. For instance, if you have a 12 Volt battery connected to a 6 Ohm resistor, 2 Amps of current will flow.

By the rules of algebra, Ohm's law can be rearranged to:

$$E = IR$$

Stated in English ... Voltage equals the product of Current times Resistance. If current is 0.5 Amps, and the resistance is 24 Ohms, then voltage is 12 Volts.

Ohms law can be arranged in a third way:

$$R = E / I$$

If you have a 12 Volt source, and 3 Amps of current are flowing, then the resistance is 4 Ohms.

Considering Ohm's law, it is apparent that the current flow in a circuit depends on both the battery voltage and the resistance of the circuit. By understanding that current increases with an increase in voltage, or a decrease in resistance, you can more easily comprehend the need for certain wire sizes, or why a light bulb designed for 12 Volts will *pop* on 24 Volts.

If resistance in a circuit gets too high, little current can flow. Dirty or corroded terminals are a high resistance which prevent sufficient current flow to operate pumps or motors. Small wire has greater resistance than larger wire, so it too impedes current flow. Small wire will even overheat if used improperly.

Resistance is often referred to as the *load* on the circuit. The lower the resistance, the greater the load. A lesser resistance allows more current to flow, so it represents a greater load on the battery.

All electrical equipment, such as lights, radios, pumps, and blowers are loads and are designed to operate at a specific voltage, usually 12 Volts. The amount of current that an appliance draws from the battery depends to a large extent on the amount of resistance that the equipment presents to the circuit. Wire resistance, of course, is added to the resistance of the device. A bilge pump with a large pumping capacity has a larger motor ... it will be a greater load, and draw more current than a smaller pump.

3.10 Determining Electrical Draw

Because wire size, as well as fuse or breaker size depends on the current draw of a device, it is important to be able to determine what the draw is.

Many electrical devices have a label that indicates the proper voltage to be connected to its terminals, and the amount of current which the device draws.

Often, manufacturers provide data sheets for their products that specify current draw. Copies can be obtained from their technical support department, as long as you know the equipment model number.

All too frequently, a device has no label, and thus no part number either. The only way to find out how much it draws is by measurement. Meters, covered in greater detail later, are available to measure voltage, current, and resistance. Direct current measurement of the device in normal operation is the best way to determine draw.

Direct current measurement isn't always possible. Voltage is most easily measured, so if you also measure the resistance of a device, you can determine current by applying Ohm's law.

There is a caveat, however. Resistance, as measured by an ohmmeter, is not always the resistance seen by the circuit in actual operation. If you measure the resistance of a light bulb, the ohmmeter will indicate a very low resistance. If you were to apply Ohm's law to the measured resistance, you'd find that the predicted current would be huge. You know that a compass light doesn't draw several Amps of current ... what's wrong?

The ohmmeter imposes a very small voltage across the bulb. It measures the current flow and determines the resistance from the voltage and current. In operation, a much larger voltage is applied to the bulb. As a consequence, the bulb's filament gets hot. As a wire gets hot, its resistance increases. In the case of a filament, the increase in resistance may be ten or more times that which is measured when the filament is cold.

Measured resistance of motors is also misleading. We have yet to explain the theory that will allow a complete description of what happens to a motor when it is operating. At this point, accept the statement that a motor generates a *back electromotive force* that makes its operating resistance greater than its measured resistance. Its actual draw will be less than calculated using the measured resistance.

3.11 Power

Voltage or current measurement alone does not indicate the amount of electrical work that is being done. To determine how much work the electricity is actually doing, both voltage and current must be known simultaneously. Power is simply the product of voltage times current. The unit of power is the Watt, abbreviated with the letter W. A 12–Volt light bulb which draws 2 Amps, consumes 24 Watts. The Watt is named in honor of James Watt[1]

Most light bulbs are labelled with their rated wattage so you can easily determine the current that a bulb will draw. Divide the wattage rating, by the applied voltage. The amount of current drawn by individual light bulbs can be useful information when installing a lighting system. In this situation, it is important that wire and circuit breakers are sized for the sum of currents for all the lights.

A simple exercise illustrates why devices must not be connected to a voltage higher than their rating. A 12–Volt pump that draws 10 Amps, consumes 120 Watts of power. If the pump is properly designed, about 70% of that power will be applied toward pumping action. The rest will be wasted in the form of heat. If the 12–Volt pump is connected to 24 Volts, it is reasonable to assume that 20 Amps of current will result. The power now consumed by the pump is 480 Watts, a fourfold increase. Whereas on 12 Volts the pump was heated by 36 Watts, (30% of 120), on 24 Volts the pump is heated by 144 Watts. Since the pump isn't designed to dissipate this much heat, it will quickly be destroyed.

You can put many 24–Volt devices in 12 Volt circuits without problems. Motors rated for 24 Volts will turn at half speed when operated on 12 Volts. Light bulbs will be 1/4 brightness. Not all 24 Volt devices can be run on 12 Volts. Fluorescent fixtures designed for 24 Volts will not run properly on 12 Volts. DC–AC inverters can not be used at lesser voltages either.

3.12 Batteries and Voltage

We've talked about batteries as voltage sources. A 12–Volt battery is not exactly a 12–Volt voltage source. More details about batteries are given later. Here, we want to point out, that the voltage on a 12–Volt battery can vary under use from almost 13 Volts, to 7 or 8 Volts.

A fully charged battery with no load attached will rest at about 12.9

[1]Scottish engineer, 1736–1819

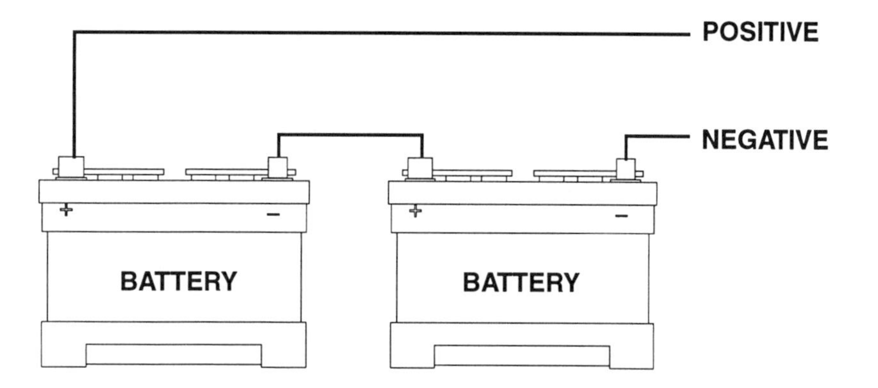

Figure 3.2: Batteries in Series

Volts. Under heavy demand of a starter, or a DC–AC inverter, the battery voltage will drop substantially. A starter load may momentarily drop the battery voltage to 7 Volts. A high powered inverter regularly lowers battery voltage to the 11 Volt vicinity. A battery should never be operated with a sustained load that takes the voltage below 10.5 Volts, and for longest life, operation should cease at 11.5 Volts for heavy draw (greater than 50 Amps). For light loads, (under 15 Amps) do not operate a battery after its terminal voltage has fallen to 12 Volts.

Voltage on a battery declines under load due to internal resistance. Internal resistance of a battery is not a constant value under all conditions of load current, and state of charge, but you can determine a reasonable value for internal resistance by measuring the voltage drop when subjected to two different load currents. Excessive internal resistance indicates a battery whose remaining life is short.

To determine internal resistance, measure the terminal voltage for two different load currents. Suppose a battery has a voltage of 12.7 Volts for a 5 Amp draw, and 12.6 Amps for a 25 Amp draw. The difference in voltage is 0.1 Volts for a difference in Amps of 20. Using Ohm's law, with 0.1 divided by 20, the internal resistance is 0.005 Ohms.

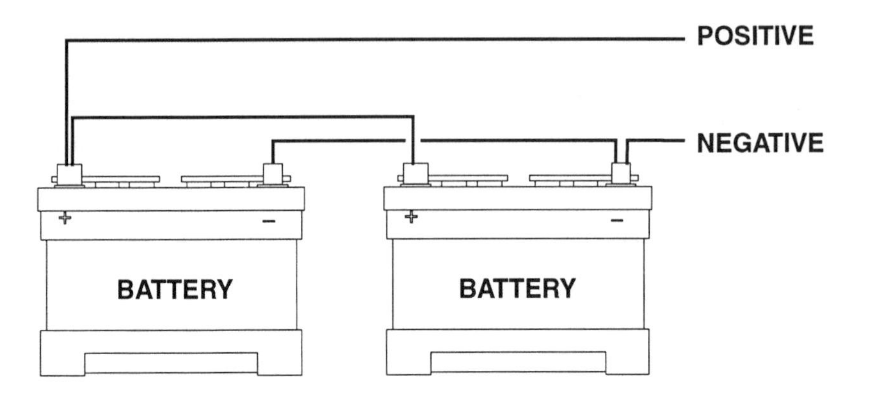

Figure 3.3: Batteries in Parallel

3.13 Series Connections

Voltage sources can be connected in series. The voltage attained by series connected sources is the sum of the total. Two 12–Volt batteries connected in series make a 24 Volt source. Figure 3.2 shows two batteries in series. Note that the positive of one battery is connected to the negative terminal of the other.

When two voltage sources are connected in series, the current rating is the lesser of the two. For two batteries in series, the total Amp hour rating is that of the smallest battery.

3.14 Batteries in Parallel

Voltage sources may be connected in parallel. Figure 3.3 shows two batteries connected in parallel. Note that positive leads are connected to each other, and negative leads are likewise connected. Only voltage sources of equal voltage should be connected in parallel.

When sources are connected in parallel, the voltage remains the same. The current rating, however, is equal to the sum of the two. If a battery of 150 Amp hours is connected in parallel with one of 300 Amp hours, the result is 450 Amp hours of capacity.

Note that the positive and negative wires are taken from opposite batteries. This technique is done to balance the resistance to each of the batteries and therefore assure an equal discharge and charge potential.

The technique can be extended to any number of batteries. Take the positive and negative wires from opposite ends of the parallel batteries.

3.15 Batteries as Loads

When a battery is being charged, it acts as a load device, rather than a voltage source. The battery charger is the voltage source in the circuit.

3.16 Capacitors in DC Circuits

We are all painfully aware that electrons can collect on our bodies in wait of the door knob. The electrons are separated from carpeting by friction. When we touch something that conducts back to the positively charged carpet, we get zapped.

A device designed to collect electrons is the capacitor, sometimes called a condenser. The capacitor consists of two metallic plates which are insulated from each other. The plates are made as large as necessary, and the closer they can be without touching, the better. The more surface area of the capacitor plates, the more storage capacity.

A capacitor does not pass DC current. When a voltage source is first attached to a capacitor, a high inrush of current results, without a corresponding increase in voltage. Eventually, the capacitor charges to the same level as the driving voltage source, and current flows stops.

If the voltage source is removed from the capacitor, the charge on the capacitor remains unless a load resistance across the capacitor discharges it. Shock hazards can exist in high voltage circuits due to stored charges on capacitors.

Capacitors are used in DC circuits to filter noise. By its actions, a capacitor opposes any change in voltage across its terminals. Voltage sources that don't produce a constant voltage can be smoothed by a capacitor. A capacitor across an alternator output, for instance, can reduce the electrical noise of the alternator. A capacitor at the input of an electronic device can prevent voltage spikes from disrupting operation.

Capacitors are made for different applications. Those designed for large storage are polarized, that is, they have a positive and a negative terminal that must be observed when connected. Capacitors designed

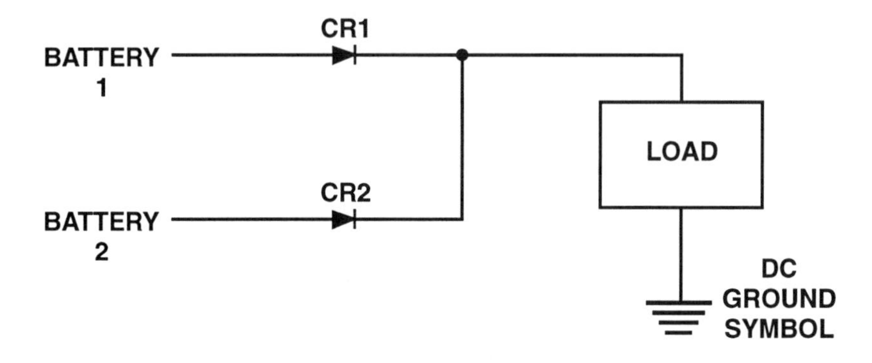

Figure 3.4: Diode 'Or' Circuit

for high frequency filtering are not polarized, and are much smaller than storage capacitors.

3.17 Diodes

Diodes are the simplest of semiconductor devices. A diode only conducts current in one direction. A diode that is conducting is said to be *forward biased*, while one that isn't is said to be *reverse biased.* Unlike a resistor that develops a voltage across it proportional to the current through it, a diode has a relatively fixed voltage drop when forward biased. The drop is from 0.5–1.5 Volts. A reverse biased diode can sustain a high voltage across it, in some cases, thousands of Volts.

Diodes have a forward current rating, and a reverse voltage rating. Exceeding either, can result in destruction. To operate at its forward current rating, a diode must be cooled, either with a fan, or a heatsink device.

Later in the chapter about AC electricity, we'll find out that diodes convert AC to DC. Diodes are also useful in DC circuits. One application for diodes in DC circuits is to form a *logical* **or** function. Figure 3.4 shows two diodes connected together at a load. Each diode is connected to a different battery, called Battery 1 and Battery 2. The load is powered when either Battery 1 **OR** Battery 2 is charged. Because diodes only conduct in one direction, the two batteries are not connected together, yet either of them can supply power to the load. Which battery supplies

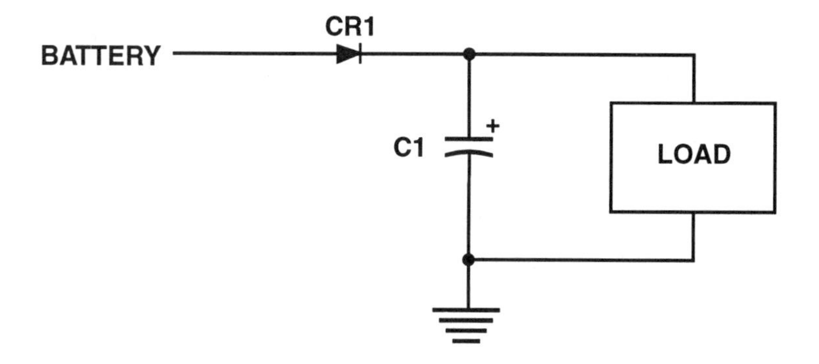

Figure 3.5: Diode and Capacitor

the power? The one with the highest voltage.

Note the symbol which we use to represent DC ground. DC ground is a logical entity which represents the negative terminal of a voltage source. The physical entity of DC ground is an actual negative distribution terminal.

Diodes can be combined with capacitors to provide *decoupling*. In Figure 3.5 a diode is used to keep a capacitor charged and a load powered. If the battery voltage were to take a sudden dip, as happens when a starter motor is engaged, the capacitor will continue to supply load current. C1 is referred to as a decoupling capacitor because momentarily, the voltage on the load is not the same as the battery.

Note that the capacitor has a plus sign at one end. Decoupling capacitors are usually polarized because it is easier to make large value capacitors polarized. Never hook up a polarized capacitor backwards, it can explode.

3.18 Summary

In this chapter, you have learned the basics of DC electricity. The rules and entities of electricity make up a formal system. Voltage, current and resistance are three basic *objects* in this formal system. The relationships between these objects are described by Ohm's law ... current increases for an increase in voltage, or a decrease in resistance. Direct current flows in one direction, from the negative terminal of a source, through the load to the positive terminal. A closed, or complete circuit is required for sustained current flow.

Operations on the objects include those which algebraically manipulate Ohm's law, as well as the power equation ... power (in Watts) is equal to the product of voltage times current.

A voltage source is required to force the flow of current. It is the flow of current that performs electrical work. Electrical work is done by generation of heat, or a magnetic field. Conductors and voltage sources are not perfect since both have resistance.

Chapter 4

DC Magnetics

4.1 Introduction

We're all familiar with magnets ... they stick notes to refrigerator doors. Magnets attract, or are attracted to iron, depending on your point of reference. The attraction operates at a distance, even penetrating cardboard. Without magnetic forces, the realm of electricity would only be useful for generation of heat. Magnetic forces allow such devices as motors and generators to operate. The transformer that allows long distance transport of AC power would not function if it were not for magnetic force.

Magnetic force did indeed seem strange to ancient Greeks, and you may even find it a little baffling. Magnetism, however, is well understood, and magnetic behavior is predictable. By learning a few simple relationships, you will begin to see how magnetism has been put to work for the immense benefit of the human race.

4.2 Permanent Magnets

The ancient Greeks noticed that a lead–colored stone from a place called Magnesia, attracted particles of iron ore. It was discovered that the stone always pointed in the same direction when it was suspended by a string, or floated on a piece of wood in water. The magical stone was called a lodestone, which meant *leading stone.* Indeed, since the stone always pointed in one direction, it was used to lead ancient explorers.

Today, we know that the earth is a giant magnet, and a suspended or floating magnet aligns itself with the magnetic poles of the earth. You may recall that like charges repel, while unlike charges attact. A similar relationship exists for magnets. We have come to call the ends of the earth as the north and south poles. Likewise, a magnet is said to have a north and south pole.

Artifical magnets can be made, and they are stronger than natural magnets. Iron, cobalt, nickel and other rare earth elements are used to make magnets. Iron is easily magnetized, but quickly loses it magnetism. Steel is harder to magnetize, but retains magnetization longer.

In a substance that can not be magnetized, the electrons orbiting the nucleus spin in opposite directions. Each atom has its own north and south pole, but the poles are arranged more or less randomly, cancelling any detectable magnetic force. In a magnet, the electrons spin in the same direction, so the small forces of each electron add up instead of cancel. Heating a magnet can destroy it, because hot atoms move faster than cold ones, and during movement tend to take on a random orientation. Striking a magnet sharply can also demagnetize it.

Magnets have invisible lines of force, which connect between the two poles. These lines of force can be made visible by sprinkling small iron filings over a magnet. If you attempt this experiment, it is best to place a piece of paper over the magnet so that the iron filings won't attach themselves to the magnet. The stronger the magnet, the more lines of force it will have per unit area.

Magnetic force is always stronger near the poles. In fact, magnetic force varies inversely with the square of the distance. That is, if you half the distance between a magnet and iron, the attraction is four times as strong. Conversely, if you double the distance, the attraction is only 1/4 as strong. Gravity obeys the same rule, but as yet, the connection between gravity and magnetism has not been reduced to scientific explanation.

Iron and steel distort magnetic fields in their presence, including the earth's magnetic field. Compasses must use compensating magnets whenever they must operate in the presence of iron or steel. The poles of compensating magnets are carefully oriented so that the small magnetic needle of a compass points correctly at any rotation relative to the earth's poles.

Magnetic fields follow familiar rules you have learned about electricity. Instead of a voltage which forces electrons to move against circuit resistance, magnets posses a *magnetomotive force.* Magnetomotive force cause magnetic *flux*, an analogy of current flow. Flux is opposed by reluctance, rather than resistance.

The amount of flux created by a magnet is given by:

$$\text{flux} = \text{magnetomotive force} \; / \; \text{reluctance}$$

Does this remind you of Ohm's law for electricity?

Flux is the number of lines of force that exist around a magnet. The unit of flux is the *Weber*, named in honor of Wilhelm Weber[1]. The concentration of flux, or *flux density* is the number of lines of force in a given area. Flux density is measured in units of *Tesla*, in honor of Nikola Tesla[2]. One Tesla is one Weber per square meter.

The term reluctance is not frequently used. Instead, the reciprocal of reluctance, *permeance* is used. Permeance is the ease with which a material can be penetrated by flux. Air has a permeability of 1. Permability for other mediums is measured relative to air. Iron has a high permeability. Flux density is the product of permeability times field intensity, so the more permeable a material, the greater the flux density.

4.3 Magnets and Electricity

The link between magnets and electrical current was established when it was discovered that a current carrying conductor deflected the needle of a compass. Current through a conductor produces magnetic lines of force that are circular, and at right angles to the current flow. These rings of force are evenly distributed along the length of the conductor.

As mentioned, permanent magnetism results when all of the electrons spin in the same direction. Each spinning electron is radiating a tiny magnetic field, and when they are aligned, the effect is magnetism. The magnetic force developed along a straight wire is generally too weak to be of much use. By winding the wire in a loop, or coil, the magnetism in a given area becomes concentrated. The intensity of the field is directly

[1] German physicist, 1804 – 1891.

[2] Croatin–American electrical engineer, 1856 – 1943.

proportional to the number of turns in the coil, the amount of current flowing, and the length of the coil. Magnetic field intensity is denoted by the letter H.

By placing a piece of soft iron inside the coil, additional concentration of flux density results. This apparatus is called an electromagnet, since the iron will develop a north and south pole exactly like a permanent magnet.

The principles of electromagnetism are used in relays and solenoids, for instance the starter solenoid. A coil of wire around a moveable iron core acts to center the iron whenever current is flowing. A starter solenoid has a coil, plus a moveable iron core, sometimes called the *coil piece.* The iron core is held off center by a spring. When you switch current to the solenoid coil, the electromagnet overpowers the spring, pulling the iron core to the center of the coil. When the iron core moves to the center of the coil, it also moves an electrical contact with it, closing the circuit to the starter motor. By using a solenoid, hundreds of Amps of current can be controlled by a relatively small amount of current that is necessary for the electromagnet coil.

The telegraph operated on the principle of electromagnetism, similar to the solenoid. Electrical relays make use of the same principles, whereby a small current is able to switch a much larger current. You should note that there is no electrical connection required between the coil, and the contacts that are closed by the moveable coil piece. This allows isolated control. For instance, a 12–Volt DC coil can be used to switch contacts of a 115 Volt AC circuit.

Mechanical motion obtained by a simple coil and coil piece can be used directly. Small engine driven generators often use a solenoid to keep the engine throttle open. The coil of the solenoid is wired in series with switches that open if high water temperature, or low oil pressure occur. If a switch opens, the solenoid looses power to its coil, and the throttle is shut off by a return spring.

4.4 Motors and Generators

Since current through a conductor exerts a magnetic force, it was natural to think of ways to obtain continuous motion using magnets and current flow. In Figure 4.1, a wire is shown between the poles of a magnet. A

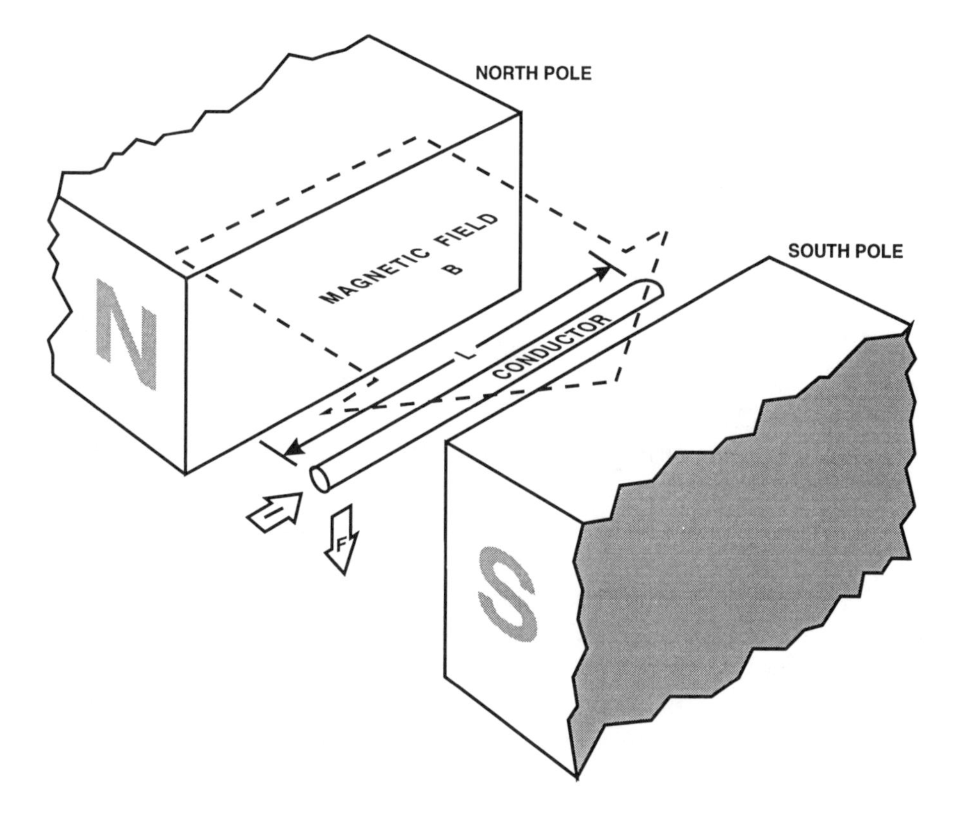

Figure 4.1: Wire in a Magnetic Field

magnetic force, B, is present between the poles. The conductor length within the magnetic field is L units long, and is carrying I Amps of current. A mechanical force, F is developed 90° to the wire. The amount of force is given by the simple equation:

$$F = BLI$$

In this equation, the units of force, F, are *Newtons*, units of magnetic force B are *Teslas*, the length L is in meters, and I is in Amps. A Newton is a mechanical force that can accelerate one kilogram, one meter per second per second.

In Figure 4.1, the wire or the magnet would be accelerated from the position shown, if one or the other was not restrained. The direction of the acceleration is dependent on the direction of the magnetic field, and the direction of the current. That's handy, because the direction of motion can be changed by simply reversing the direction of current flow.

While the electromagnetic laws were not yet worked out, Faraday was able to make a motor, which provided continuous rotary motion. Figure 4.2 shows the type of motor constructed by Faraday. A conductive disc is placed in a magnetic field, such that it can rotate freely. Two sliding contacts, called brushes, supply current to the disc. Since the current will flow between the brushes, you can think of the conductor shown earlier in Figure 4.1 as a length of wire between the brushes. This conductor will be forced to move by the resultant force. Since the disc is restrained by its axle from moving out of the magnetic field, it will rotate instead.

The Faraday motor is called a single pole, or homopolar motor. It operates on low voltages with high currents necessary. In practice today, motors have several poles, and rather than a short conductor on a disc, the conductors are wound coils which operate with higher voltages, and less current. Brushes today are usually made of carbon.

A generator is the same apparatus as a motor. In the case of a Faraday generator, the axle of the disc would be driven by an external force, and current would be generated between the brushes. All modern generators work on the same basic principles.

In the chapter on DC Electricity, we mentioned that a motor develops a *back electromotive force* that makes its operating resistance appear different than that measured with an ohmmeter. Now that you have learned that a motor and generator are essentially the same machine, the term

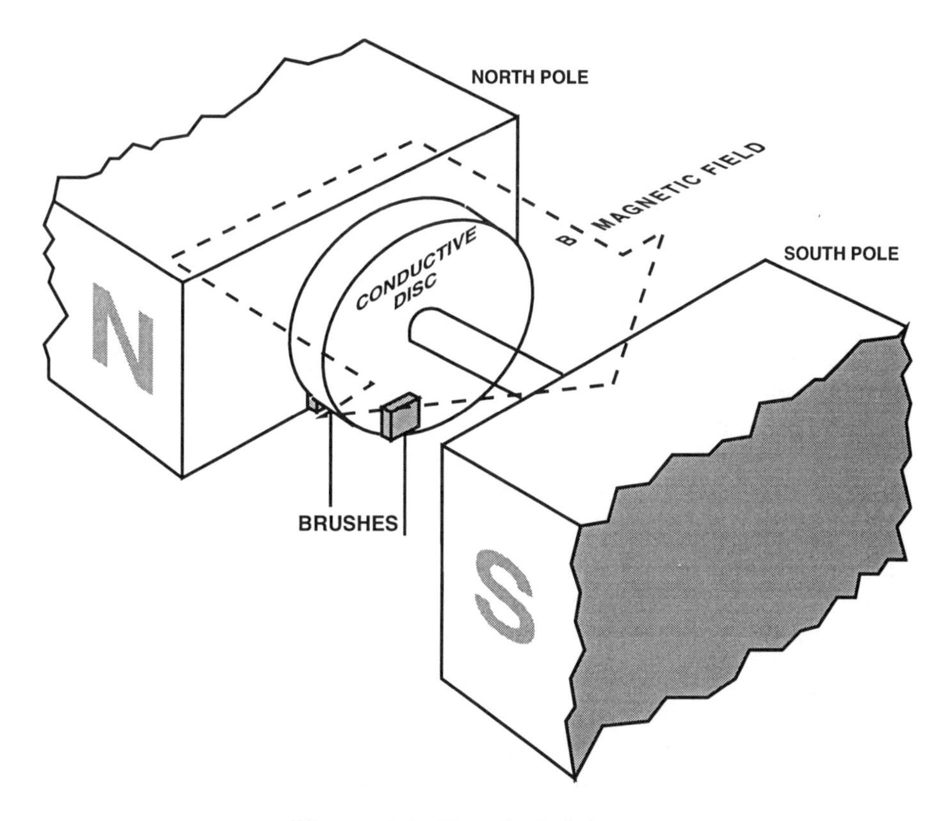

Figure 4.2: Faraday's Motor

back emf can be explained. As a motor is driven by current through its conductors, the magnetic field that it is spinning through wants to also generate electricity in the conductors. The generated electricity flows in the opposite direction than the motor current. The voltage source for the motor current has to *buck* the generated current. The actual current through the conductors is therefore less than it would be if the motor winding were a simple resistance. An efficient motor design seeks to reduce back emf to a minimal level, but it can't be completely eliminated.

4.5 Inductance

A coil of wire has a property called inductance. If you measure the resistance of a coil, you will find that it is quite low. An alternator field coil, for instance has a resistance on the order of several Ohms. Suppose a field coil has a resistance of 2 Ohms. If we apply 12 Volts to the coil, we expect that 6 Amps of current will flow. Eventually, 6 Amps will flow, but not the instant that voltage is applied. The delay between connecting the voltage and obtaining 6 Amps is caused by inductance.

As mentioned, a current carrying conductor develops a magnetic field around it. A wire which moves through a magnetic field has a current induced in it, which is the basis of the generator. The wire doesn't have to move through the magnetic field ...the field can move through the wire.

When voltage is first applied to a coil, a magnetic field begins to develop around the coil. As the field develops, it moves through the wires of the coil, inducing a current in them. As in the case of back emf, the induced current is in the opposite direction of the applied current. The net effect is a reduced current flow for a period after the coil first has voltage applied.

In power electrical circuits, inductance is seldom a problem when power is first applied. Instead, it is often an advantage. At the instant switch contacts close, inductance prevents an inrush of current. This gives the contacts time to close firmly on their full surface area before spot heating can burn parts of the contacts. Inductance rears its ugly head when a switch is opened.

When power is removed from a coil, the magnetic field which was produced by the current can no longer exist. The magnetic field is said to collapse. The collapsing magnetic field moves through the wires, and in doing so works to induce current. The induced current flows in the same direction as the applied current was flowing just before interruption. But, you have opened the switch to the coil, so the current has nowhere to go. The voltage rises instead, and the coil becomes a high voltage generator. The high voltage arcs across the switch contacts, burning them. This phenomena explains why a capacitor, or condenser, is used across automotive distributor contacts. The condenser absorbs the high voltage spike that results when the points are opened, thus preventing arcs.

Switch contacts don't usually fail immediately from an *inductive kick*, but over time, repeated arcing takes it toll. Placing capacitors in parallel with coils (inductors) means that on initial switch closure, there will be an inrush of current to charge the capacitor. This inrush can arc across contacts that are not yet firmly closed.

Switch contacts can also be protected by diodes wired in parallel with a coil. Figure 4.3 shows a foot switch for an anchor windlass. The foot switch operates a large solenoid which in turn supplies power to the windlass. The foot switch is protected by the diode. Note that the diode is connected such that it does not conduct when the foot switch is actuated. When the foot switch is released, the inductance of the solenoid turns it into a generator which attempts to maintain the same magnitude and direction of current that existed prior to the foot switch opening. The polarity of the generator is such that the diode conducts, allowing the current through the coil to decay safely.

Switches connected to starter solenoids, electric fuel pumps, and even pumps can be protected by diodes. The slight cost of doing so can prevent a switch failure at an inconvenient time or location. Any time a semiconductor is used to switch current on and off to a coil, a protection diode is mandatory. A 1 Amp silicon diode is an inexpensive bit of insurance.

4.6 Summary

In this chapter we have dealt with the subject of DC magnetism. Permanent magnets have electron spin aligned so that the tiny magnetic field of each electron is additive. Electricity produces magnetic fields due to current flow ... electromagnetism. Solenoids and relays are made with a

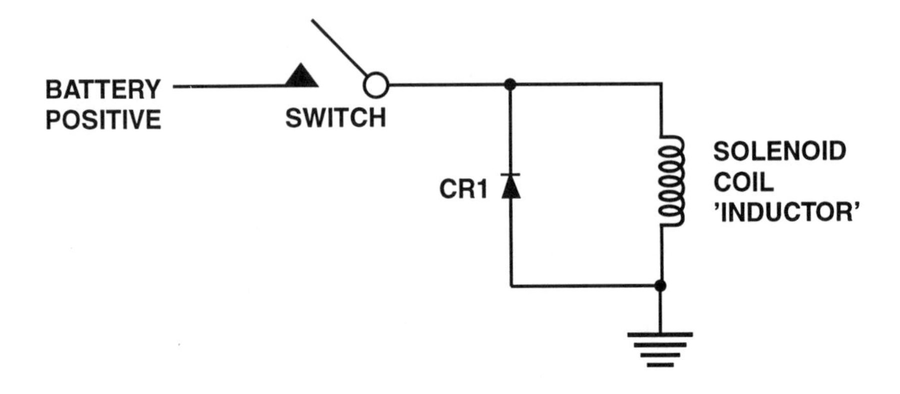

Figure 4.3: Diode Protecting a Switch

coil of wire which moves an iron rod when current flows.

Current flow within a magnetic field results in a mechanical force acting between the magnet and the current carrying conductor. A motor operates on these principles. A magnetic field which moves through a wire induces current flow, which is the basis for the generator.

Coils have inductance, an attribute that tends to oppose a change in current flow. Inductance can create high voltages and arc across switch contacts. Capacitors or diodes can be used to prevent switch arcing.

Chapter 5

AC Electricity

5.1 Introduction

If you have understood the preceding chapters about DC electricity, and magnetism, then understanding AC electricity will not be hard. AC electricity can not be understood, however, without roots in DC theory.

AC stands for **A**lternating **C**urrent. Instead of flowing continually in one direction, alternating current flows for a time in one direction, and then reverses direction for an equal length of time.

A familiar device, the transformer is introduced in this chapter. Transformers don't operate on DC. Characteristics of the capacitor and inductor are revisited. Whereas a capacitor looks like an open circuit to DC, it offers a low impedance to AC. The inductor, which is a low resistance to DC represents a high impedance to AC.

Diodes in AC circuits are used to convert AC to DC, a process called rectification.

5.2 Alternating Current

As mentioned, AC current does not continually flow in one direction. In the AC circuit, current flows in one direction for a short period, and then reverses. The current is of equal amount in both directions. AC current is said to have a frequency ... the number of complete cycles in one second. In the U.S., AC frequency is 60 times per second, or 60 Hertz. The unit

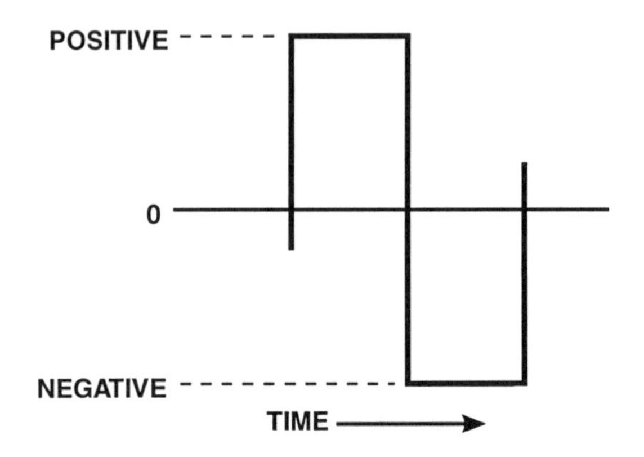

Figure 5.1: Square Wave Current

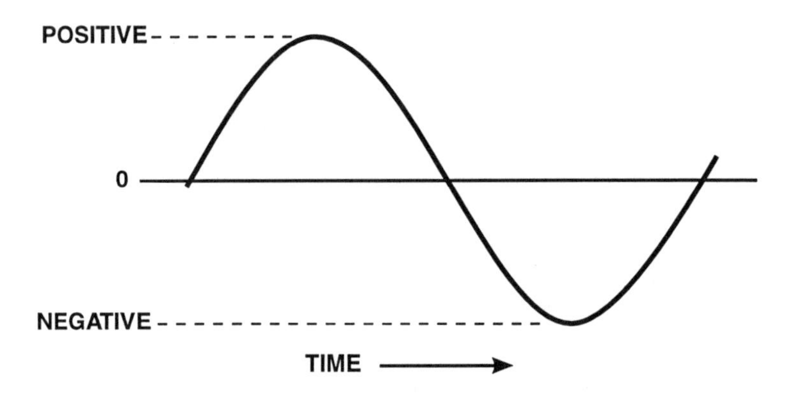

Figure 5.2: Sine Wave Current

of frequency is Hertz, in honor of Heinrich Hertz[1].

Normal household AC current does not switch immediately from high current in one direction to high current in the other direction. Such a switch would generate a square wave. A square wave is plotted against time in Figure 5.1. Current in the top half of the figure is called positive current, while current in the bottom half is called negative current.

Instead of switching immediately from high positive current to high negative current, household AC goes through the transition gradually. Figure 5.2 shows a typical sine wave of AC current. The current is a

[1]German physicist, 1857 – 1894.

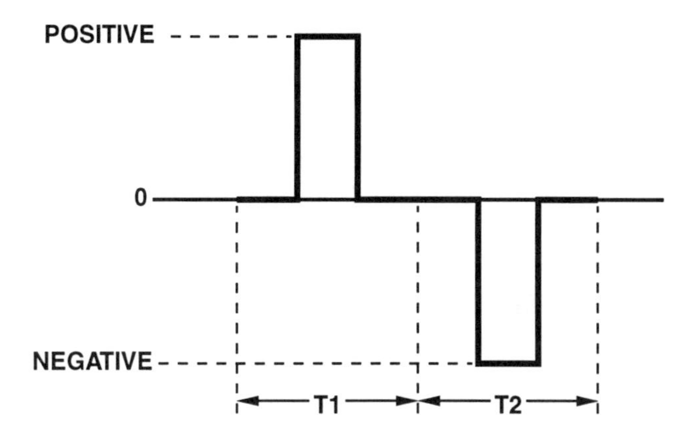

Figure 5.3: Inverter Modified Square Wave

sine wave because of the nature of generators. As the generator rotates a magnet inside conductors, the distance from the poles of the magnet to the conductors in the generator sweep out a sine wave. The distance varies with time according to Figure 5.2. We learned earlier that magnetic force varies with distance, so it follows that current from the generator will be greater or less, depending on the distance between the rotating magnet and the stationary conductors.

Current reverses in the generator because the rotating magnet first presents one pole to the conductors, and then the other pole. The frequency of current reversal is the result of the rotating RPM of the generator, which explains why auxiliary generators must have a governor. Poorly controlled generators do not maintain a constant frequency, and electrical problems can consequently occur.

Many AC appliances are not affected by frequency. Toasters, lights and other devices that work strictly on heat will work on DC as well as AC. Some motors, called universal motors, will also work on DC. Many motors, however, will only work on AC. Motors designed to operate on 60 Hertz, will usually work on 50 Hertz, but will not be able to drive as big a load.

Ferroresonant battery chargers are extremely sensitive to frequency. As you will learn later, there is nothing good that can be said for ferroresonant chargers.

A device called an inverter converts DC to AC. Some inverters produce the square wave shown in Figure 5.1. Most inverters produce a modified version of the square wave, and call it a modified sine wave. The wave shape produced by efficient inverters is shown in Figure 5.3. The wave shape differs from a square wave in that the width of current is varied depending on the load. High current loads will have a wider current pulse applied to them. The waveshape of an inverter that varies the width, is called a rectangular wave. In Figure 5.3 T1 and T2 show the equivalent times for either a sine or square wave.

5.3 RMS Value

For DC current, we have Ohm's law and the power equation to tell us vital information. Power, in Watts, is simply the product of Volts times Amps, and it doesn't make any difference which way the current is flowing. If AC current were a square wave, we could still use the same power equation to determine the amount of electrical work performed. We'd simply take the peak voltage, and multiply by the current at the peaks to attains Watts. In doing so, we'd be ignoring the short time required for current reversal.

Because current is not constant throughout the sine wave cycle, we can not simply take the peak AC voltage times the peak AC current to determine the work performed. What operation gives equivalence between an AC sine wave and a DC voltage? An operation called *root means squared*, or RMS. The RMS value of a sine wave is 0.707 times its peak. When you hear about 120 Volts of AC, it is the RMS value that is 120 Volts. The peak AC voltage is 120 divided by 0.707 or about 170 Volts.

The RMS value of a rectangular wave can be calculated using algebra. The steps require you to square the voltage of the rectangular wave, compute the average voltage of the wave by its on–to–off ratio, and then take the square root of that average value. The derivation of the RMS value for a sine wave is a fairly complex exercise that involves some calculus. It's easier to remember that the RMS value of a sine wave is 0.707 times its peak voltage.

We might forget all about RMS values except that most instruments for reading AC voltage and current actually read an average value instead of an RMS value. For a sine wave, it is easy enough to calibrate an average reading to an RMS value, but when the AC voltmeter or ammeter

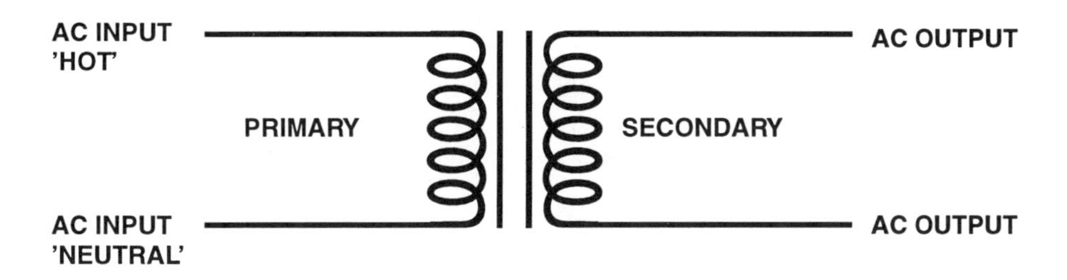

Figure 5.4: Transformer Schematic

is connected to an inverter, erroneous readings result. If you have AC instrumentation, you probably know by now that it doesn't give correct answers for the inverter. If you want correct readings, you will have to get RMS reading instruments.

AC power is often specified in VA instead of Watts. The symbol VA stands for the product of Volts times Amps, which is equivalent to Watts. Because AC is not a static voltage and current like DC, it is possible to have phase differences between the peak voltage and current. The phase differences are caused by reactive loads such as electric motors which have large inductances. If there is a phase difference between peak voltage and current, the instantaneous product of Volts and Amps will be less than it would be without the phase difference. The instantaneous product is called apparent power. The generator must still deliver real power. The ratio of apparent power to real power is called *power factor*. A power factor of 1, indicates that no phase difference exists between peak voltage and current.

Industrial plants with many electric motors must correct power factor so that the plants pay for the power they use. Correction of power factor for a motor is done by adding parallel capacitors. For small generators, a larger motor can be operated if the power factor is corrected to be close to 1.

5.4 Transformers

Transformers are devices that couple an AC voltage without a direct connection. Transformers can not only couple a voltage, but they can step a voltage up or down. That is, you can input a voltage of 220 Volts to a transformer, and get 120 Volts out.

While you can convert voltages with transformers, you can not get out more power than you put into the transformer. Transformers are 95–99% efficient, so there is not much power lost. Like other electrical apparatus, transformers have ratings. If you use a transformer beyond its rating, expect it to die of overheating.

Transformers do not operate on DC current, only AC. Recall that as current begins to flow in a conductor, a magnetic field is produced. The magnetic field originates at the conductor, and radiates outward in circles around the conductor. Recall also that a moving magnetic field which permeates a conductor, induces a current in that conductor. These facts are the basis of a transformer. Because AC is a varying voltage, its application to a coil of wire produces a varying, or *moving* magnetic field.

A transformer consists of a primary winding and one or more secondary windings. The windings are coils of wire, which are tightly wound, similar to a solenoid. The primary and secondary windings are wound on a common iron or steel *core*, that is highly permeable. The AC voltage applied to the primary winding is coupled to the secondary windings. As current in the primary winding alternates, it produces an alternating magnetic field. The alternating field induces an alternating current in the secondary windings. Figure 5.4 is a schematic diagram for a transformer.

The windings of a transformer are wound around a steel core with high permeability. Generally, the core is made up of 30–40 stamped metal pieces that are stacked together. This type of core is called a laminated core. The laminations tend to vibrate at 60 Hz, causing an audible hum.

A new type of transformer is wound on a toroidal core that looks quite like a donut. The toroidal core may be cast, or tape wound with many very thin steel bands. Toriodal transformers don't hum, and they weigh about half of what the laminated transformer weighs, Watt for Watt. Yes, the toroidal transformer does cost more.

Transformers allow the transport of electricity long distance. If you review Ohm's law, it can be seen that high voltages can develop more power across large resistances than can low voltages. Since all conductors have resistance, it follows that high voltages are more efficient for long conductors where the resistance can be significant.

Public utilities use transformers to step up the voltage produced at the generator so it can be sent long distances over power lines. At the site of delivery, another transformer steps the voltage back down so that the consumer can use electricity without unreasonable shock hazard.

A transformer couples voltage from primary to secondary according to the ratio of the number of turns in each winding. For example, a transformer with 100 turns in the primary and only 50 in the secondary is a step down transformer. If 220 VAC is applied to its primary, the secondary will yield 1/2 of the input, or 110 VAC.

For many transformers, it doesn't make any difference which winding is the primary and which is the secondary. A tranformer with 100 windings and 50 windings can be used as a step–up or a step–down transformer, depending on which way you wire it.

Transformers provide isolation, that is, there is not direct connection between primary and secondary. Sometimes transformers are used strictly for isolation reasons, since shock hazards can be minimized. Medical instruments, for instance, use specially shielded transformers so that patients connected to diagnostic equipment don't get electrocuted. Isolation transformers on boats are useful to reduce the risk of electrolysis. Indeed, any AC operated appliance aboard a boat that also connects to the DC supply should be isolated. Appliances such as toasters and hair dryers do not contain an isolation transformer, but they are not normally connected to the DC supply, so they present no electrolysis problems.

AC operated water heaters can be a source of electrolysis. The AC green wire is connected to the case of the water heater, and generally the water being heated is connected to the DC ground. Without an isolation transformer, any leakage between the AC inputs and the AC green wire can be conducted to the DC ground. Zincs and other hardware can disappear as a result.

5.5 Capacitors

Capacitors oppose a change in voltage across them, soaking up electrons without a corresponding rise in voltage. At some point, of course, the capacitor can no longer sink electrons, so the voltage at its terminals must change.

Capacitors were described in connection with DC electricity. Capacitors do not conduct DC current. On DC, the capacitor charges to the level of the driving voltage. With AC, the capacitor charges and then discharges at the AC frequency.

On DC, the capacitor is an open circuit. On AC, however, the capacitor may appear to be a short circuit, depending on the frequency, and the size of capacitor. A capacitor is said to have an impedance, or capacitive reactance, as opposed to a resistance. The impedance, X_c, of a capacitor is given by the equation:

$$X_c = 1/2\Pi FC$$

In this equation, Π is the usual 3.14159, F is the frequency in Hz, and C is the capacitance in Farads. Knowing how to compute the impedance of a capacitor can be useful if you are trying to filter noise from a circuit. The idea is to select a capacitor value which is a very low impedance at the noise frequency. Impedance can be substituted in Ohm's law in place of resistance.

Polarized capacitors can not be used in AC circuits unless two capacitors are placed in series, with like poles connected. AC changes polarity, and if a polarized capacitor is connected in reverse, it may well explode.

Capacitors suitable for AC circuits are made of ceramic, mica, or metalized films such as mylar or polypropolene.

5.6 Inductors

Inductors are coils of wire, often wound around permeable cores. The core may be a straight rod, or it may be in the shape of a toroid or many other shapes that provide useful magnetic functions.

Inductors oppose a change in current. In this regard, they are a mirror to the capacitor which opposes a change in voltage. Inductors are used with capacitors to form very effective filters.

An inductor has inductive reactance, or impedance. The impedance, X_l, of an inductor is given by the equation:

$$X_l = 2\Pi FL$$

In this equation, the symbol L represents inductance and is stated in units of Henry[2]. The greater the frequency, the greater the impedance.

[2] Joseph Henry, American physicist, 1797–1878

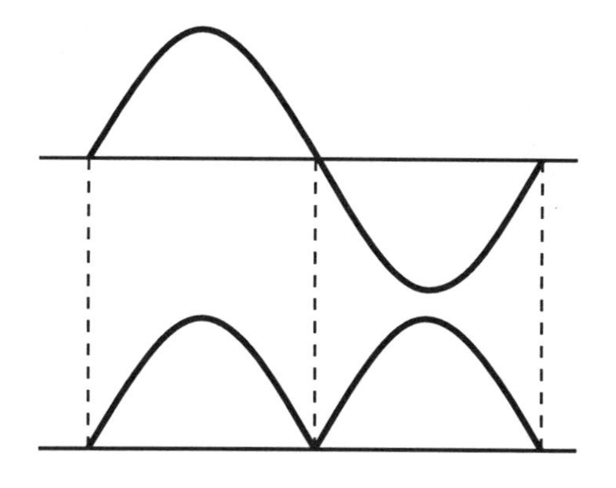

Figure 5.5: Wave Shapes Of Rectifying AC to Make DC

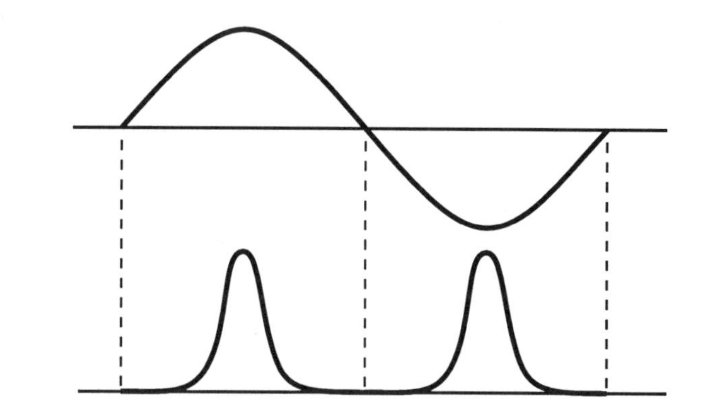

Figure 5.6: Current Wave Shape Charging Batteries

Small inductors can be placed in series with power leads to electronics. With a capacitor placed directly across the electronic equipment, little, if any, noise will reach the equipment.

5.7 Rectification

As mentioned, diodes conduct electricity in only one direction. When connected in series with an AC conductor, the diode passes current for 1/2 of the cycle, and blocks current for the other half cycle. The result is DC current. Since the DC current does not flow continuously, the DC is said to *pulse*. Figure 5.5 shows a sine wave (top) and the resultant DC

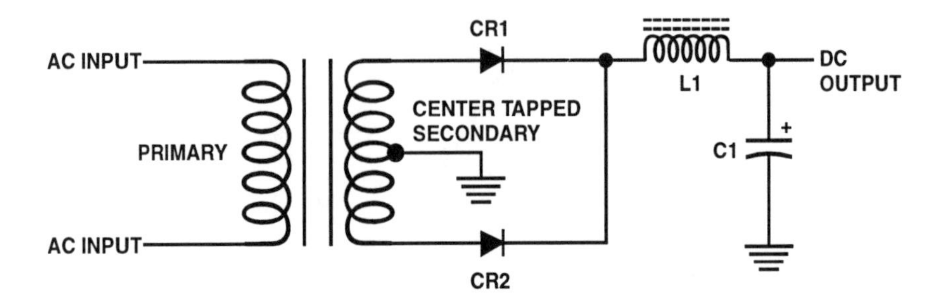

Figure 5.7: Transformer and Rectifiers

wave shape (bottom) that occurs when AC is rectified. When rectification occurs into a battery, the DC pulses shown in Figure 5.5 are filtered. The current flow that results is not filtered, in fact, its wave shape is altered significantly. Rectified current into the battery, relative to a sine wave, is shown in Figure 5.6. The battery is forced to average the pulses, which doesn't do the battery any good.

A transformer with a center tapped secondary can be *full–wave* rectified. A schematic of such a circuit is shown in Figure 5.7. One diode conducts on each half cycle of the AC sine wave.

While full–wave rectified DC is better than half–wave, it is still not sufficiently smooth to run most electronic equipment. Even batteries can not remove all the noise that comes from battery chargers that simply rectify without additional filtering.

Shown in Figure 5.7, following the diodes, is an inductor and a capacitor. The two form a filter that will smooth the pulses of DC. With an appropriately sized inductor and capacitor, the noise can be virtually eliminated. Additional inductors and capacitors can be used to provide pure DC so that the battery is not forced to act as a filter.

5.8 Summary

AC is alternating current that swings both positive and negative. Household AC is a sine wave, while DC–AC inverters produce either square waves or rectangular waves.

Transformers are devices that allow isolation of AC for prevention of electrolysis and shock hazards. Transformers, which only operate on AC, are also used to step–up or step–down voltages.

Inductors and capacitors operate differently in AC circuits. Both have an impedance which is frequency dependent.

Diodes are used to rectify AC, producing DC.

Chapter 6

Battery Loads

6.1 General Information

Battery loads are anything that consumes electricity, from lights to stereos to fans. Not all similar loads are created equal. Fans and pumps come in a variety of efficiencies depending on the quality of the motor. Lights which produce the same lighting effect do not necessarily draw the same amount of current. There may be big differences in current consumption between brands of radar. Autopilots are a special case because a mismatch between an autopilot and boat can cause a normally low power unit to consume high power. With refrigeration being the biggest consumer of electricity, special attention must be made to the type of refrigeration selected, and the amount of insulation provided.

To enjoy ample power, you will need to choose loads wisely. Don't rely on stock equipment from boat or RV manufacturers ...they have a carefully honed practice of installing the cheapest gear available, rather than equipment that makes energy sense in the long run.

6.2 Indoor Lights

There are three choices of indoor lights, incandescent, fluorescent, and halogen. The halogen light is usually thought of as strictly an outdoor spotlight, but indoor lighting fixtures are made that use halogen bulbs.

Besides the lights themselves, the light fixture optics have a significant effect on the amount of available light that is useful. Poor optics can diffuse the light so that very little is directed to the visual area.

6.2.1 Incandescent Lights

Most boats and RVs come stock with incandescent lights. Incandescent bulbs produce light by getting hot. Inside the bulb a tungsten filament glows white hot, producing light. Making an element hot just to get it to glow is a grossly inefficient process. As you might expect, most factory–built boats and RVs come equipped with incandescent lights because they are the lowest cost lighting available. We've seen some of the larger boats and RVs that can draw as much as 125 Amps if all the incandescent lights in the boat are on. Such boats are usually equipped with a single 8D house battery, and a ferroresonant charger that couldn't produce 15 Amps, flat out.

Initially, incandescent lights are cheap because the light fixture only needs to accept a light bulb and supply two prongs to conduct electricity through the bulb's filament.

We don't think that an incandescent bulb has many suitable places aboard an energy efficient yacht or RV. We have limited their use to engine instruments that are only on when the engine is running, and a compass light that is painted red and has a resistor in series to limit current.

Some make the claim that the yellowish glow of an incandescent light is so soothing that it's worth the energy draw. If you feel that way about incandescent lighting, be prepared to have big batteries, and excellent recharge equipment.

6.2.2 Fluorescent Lighting

Fluorescent lights are as much as 4 times more efficient than incandescents. Fluorescent light tubes are filled with an inert gas. When the gas has a high voltage applied across the gas tube, the gas ionizes and thus glows. Because ionization is a phenomenon that doesn't require heat, fluorescent lights are very efficient.

One drawback to a fluorescent light is the need for a high voltage to ionize the gas. To derive the high voltage, an inverter is used which converts 12 or 24 VDC to a high voltage AC ...from 60–400 Volts. The

DC–AC inverter operates anywhere from a few thousand Hz to 50,000 Hz.

Naturally, the DC to AC inverter is costly, so fluorescent lights initially cost more. Even so, with careful shopping, low cost fluorescent lights can be found. RV outlets are a good source for the most inexpensive types, but they tend to fail frequently, and generate copious radio interference. The cheaper units will often fail if you power them when no light tube is installed. Similiarly in an inexpensive unit, the failure of a light tube can cause the inverter to fail.

Optics are all important with the fluorescent lights, as with any lights. Transluscent white lens can reduce light from the tube by more than 50%. Clear acrylic prismatic lens reduce light output by less than 10%, and can provide wide angle illumination which makes for more even light distribution. The wider the distribution, the fewer fixtures required, and the less the overall power consumption.

You also have to be careful about the specifications of fluorescent lights. Just because the tube in a light is rated at 16 Watts, don't expect to get 16 Watts of light. The inverter more than likely is not powering the light tube at rated power. When you take into account inverter losses, a typical fluorescent light may only produce half of the power expected.

When choosing fluorescent lighting, 12–15 Watts is sufficient for a small cabin, or to light up a dining table. Such units are available in a single tube. For extensive reading, or closeup work such as sewing or cooking, a dual tube 30 Watt fixture is more appropriate.

Most fluorescent lights emit a white light. Some find this too bright for reading, but if an area in not over powered with light, reading under fluorescent lighting is not objectionable. Color of some tubes is less white than others.

As mentioned, inverter reliability in cheaper fixtures can be a problem. Carrying a spare fixture is advised, if you elect to use lights made for the RV market. Fluorescent fixtures have at least one other problem ... radio interference emissions. A fluorescent light can easily knock out Loran signals, and short wave weather broadcasts. Some alternative lighting may be required, such as a halogen lamp, but often, special electrical filters for interference can be installed.

Some manufacturers do produce fluorescent lighting with metal enclosed inverters that prevent most radio interference. A vendor that pro-

Figure 6.1: Alpenglow fluorescent Light

duces such high quality fluorescent lights is Alpenglow[1]. An Alpenglow light is shown in Figure 6.1. Their lights are housed in real wood of your choice, with high quality optics, and a very efficient inverter that is fully enclosed in metal to protect against electromagnetic interference. The inverters have a stable operating frequency which helps prevent noise in AM radios. FM radios are not affected, and Loran performance is normally not degraded. Alpenglow lights provide a switch to select high or low power, and they have a red light for night use. The light tubes used by Alpenglow produce as much as 30% more light for the same power input as less expensive lights.

6.2.3 Halogen Lights

In the category for efficiency, the halogen light ranks second behind fluorescent lights. Though basically an incandescent light, the halogen bulb is filled with halogen gas which allows a hotter filament. The light from a halogen lamp is generally very white.

Though not as efficient as fluorescent lights, halogen lights are better than incandescent bulbs, and do not emit the electrical noise that the many fluorescents do. At times when radio or Loran reception is critical, a halogen light can be used. At other times, the fluorescent makes the

[1]Alpenglow, P.O. Box 415, Eureka, MT 59917.

most light for the Amp draw.

6.3 Running Lights

Running lights must pass certain Coast Guard specifications. These are not necessarily very stringent. Even with the same bulb, there can be a great difference in the actual amount of light emitted through the lens, since optic quality of the lens varies significantly. A higher priced light will generally have better lens quality and thus produce more visible light at a distance. You can compare light outputs with a standard photographic light meter.

6.4 Pumps and Motors

The most energy efficient DC motors are permanent magnet types. Generally, small motors are of a wound field type that uses a separate winding to generate a magnetic field.

While pumps only run for short periods, careful attention to current draw can pay dividends.

6.5 Refrigeration

6.5.1 General Information

DC refrigeration is the single biggest user of electrical energy within the battery powered system. Particular attention must be given to conservation and appropriate use of the refrigeration system. Several means of developing refrigeration are presented below.

The most common method of cooling is by the process of evaporation. Evaporation is the process of a liquid turning into a gas. The evaporation process requires heat, which is extracted from surroundings. Evaporation is the principle behind the desert water bag, as well as the modern home refrigerator.

In the refrigerator, several components are connected to provide a continuous evaporation/condensing cycle. A pump is used to compress refrigerant gas to high pressure. The high pressure gas is then cooled by a condenser. The gas becomes liquid under the high pressure and lower temperature. The liquid is then allowed to circulate through the evaporator where it reverts to a gas and cools the evaporator. The evaporator in the home refrigerator is made of aluminum plate with tubing which transports the gas. The small DC compressor refrigeration systems which are popular in the marine market also use aluminum evaporators.

When a constant source of power is not available to drive the compressor on demand of the thermostat, holdover plates are used. Holdover plates are filled with a brine solution. The brine is frozen when refrigerant is pumped through tubes in the plates. The refrigerant enters the plates as liquid and evaporates to a gas inside the plate ...the typical evaporation process. During times when the compressor is not running, the frozen brine in the plates cools the refrigerator. Brines with different freezing temperatures are used for refrigerator or freezer units. In effect, a holdover plate acts as a *cold* battery, storing cold for later use. It even acts much like a battery when *charged* or pumped down. Initially, the plate accepts a rapid cool down, but as ice begins to form on the internal tubing, the rate of cool down tapers off. While many proponents of holdover plates suggest that one pump down per day of about 20 minutes is sufficient, such is not the case. Full pump downs require 45 minutes to an hour on any reasonable size plate. How long a holdover plate will maintain a cold box is dependent on the amount of brine in the plate. The larger the volume, the longer the plate will keep things cold. As you might expect, it takes a large volume to maintain cool temperature for a whole day. Some holdover plate systems require 2–3 pump downs per day ... a sign of too small a plate, or too little insulation.

Another method of extracting heat is the modern thermoelectric module. In this device electric current is passed through a semiconductor junction of dissimilar alloys. As current flows, one side of the junction gets hot, while the other side gets cold. By externally cooling the hot side with a fan, heat can be extracted from the box. Thermoelectric refrigeration may have some limited application in a battery powered system, but the modules consume about 3–10 times the electricity of a compressor system, given the same amount of cooling.

Absorption refrigeration is yet another method of cooling. In this technique, heat is used to separate ammonia which has been absorbed in water. The freed ammonia gas is then cooled in a condenser and later allowed to evaporate. As the liquid evaporates, the gas is recaptured in the absorbent water, where it can again be removed by application of heat. Absorption systems are not as efficient as compressor systems, but can be simpler. They operate without moving parts such as the compressor, and only require heat as an input. Heat is usually supplied by electricity or gas, although kerosene burning units are available. Continuously operating units must be maintained near level, limiting their use to RVs or remote homes.

For the most part, small marine refrigeration systems will be operated from DC or an engine. Larger system are often driven by AC which is derived from an onboard generator. Sometimes AC refrigeration is driven from a DC–AC inverter, where the DC is derived from batteries and a large engine alternator. There are pros and cons to each of these methods as presented below.

This section on refrigeration is concluded with a few words about insulation. It is only possible in this book to present a few of the most salient details regarding refrigeration. For detailed information about refrigeration, refer to the book, *Living on 12 Volts with Ample Power.*

6.5.2 Engine Driven Refrigeration

Since DC refrigeration draws so much battery power, you may think that engine driven refrigeration is the way to go. In many cases, engine driven holdover plate systems perform sufficiently well. In many other cases, the systems either don't work at all, or require too many engine operations during the day, making the owner a slave to the icebox.

Besides being tied to the engine on a regular basis, engine driven systems have other problems. One is a matter of design. Typically, an automotive air conditioning compressor is used to circulate refrigerant. This type of compressor was not designed to operate at the low pressures of holding plate evaporators. With the low pressures, undue stress is placed on the compressor seals. Even without low pressures, seals are not all that reliable. The seals are actually metal and require a good coat of oil to seal effectively. If the unit is not operated frequently (every two weeks), then the seals dry up and leak all the refrigeration into the atmosphere. Lost refrigerants are a major contributor to the atmospheric hole in the ozone layer. If you do have engine driven refrigeration or air conditioning, it will save repairs and the atmosphere if you run the compressor every two weeks even when not actively using the system.

The engine driven compressor is called to operate over a wide range of RPM. While the air conditioner compressor has been designed to operate over the required RPM range, normal open frame compressors that can withstand low pressures will not.

Another problem with the engine driven system is the use of a large engine to drive a small load. The compressor amounts to an initial load of about 2 HP, but as the tubes in the holdover plates start to freeze, the load declines. Soon, the engine has virtually no load on it. This is not good for any engine, particularly diesels. Idling diesels builds up carbon in the valves, and without full combustion pressures and temperatures, the pistons tend to rattle and the rings don't seal properly ... all detrimental effects. As noted earlier, holdover plate systems don't pump down as fast as we wish they would, so engine run time can be considerable, especially in warm climates where the engine may have to be run as much as four times a day.

A problem with holding plate systems in general is good instrumentation that can be used to determine when the plates are frozen. If the compressor is run too long, then intake pressures on the compressor will get so low that the seals will fail. This situation is aggravated when a large compressor is driving small holdover plates ... the typical case. For this reason, many have started to use timers which deactivate the compressor after a given time. Such a control scheme may save the compressor, but the plates may not be sufficiently frozen to last very long before another pumpdown is required.

What with being a slave to the engine, a good chance of losing refrigerant into the atmosphere, and unnecessary wear on the engine, we are not in favor of main engine driven refrigeration. When a small engine is used, which may be driving enough other equipment that it is operating with a considerable load, then the system makes more sense. In this situation, the engine and compressor speed can become constant so that a conventional open frame compressor may be used. Such a compressor can tolerate the low pressures without seal failure. Great amounts of fuel are not wasted in the process. You will still be a slave to the engine, of course, but the other drawbacks can be eliminated.

6.5.3 Direct DC Compressor Refrigeration

With enough battery capacity, direct DC refrigeration has advantages. It can run anytime the thermostat clicks on, so a very constant box temperature can be maintained. Such systems generally use small hermetically sealed compressors, so loss of refrigerant into the atmosphere is minimal, occurring only at service.

The small DC systems usually employ an air cooled condenser. Air is circulated through the condenser by a small fan. Air cooled condensers are not very efficient in warm weather, and in hot weather, they are practically worthless. Problems with air cooling are more often than not aggravated by a condenser mounting location that has no fresh air to circulate. We've seen systems mounted in small enclosed areas that heat up faster than the box cools down. The net result is alot of electricity consumed, but little refrigeration to show for it.

If you plan to cruise in the tropics, with a direct DC system, you must install a water cooled condenser. While current consumption will increase due to running a water pump, the rewards of more efficient condenser operation will more than offset the additional load.

Plan on more battery power than the refrigeration manufacturer suggests. They tend to focus on the benefits of refrigeration, not the reality of powering it. Despite claims otherwise, DC refrigeration is a huge load on the batteries and charging system. Current draw is always more than specified on the refrigeration brochures which often assume that the unit will operate only 20% of the time. In practice, boxes are too large with too little insulation for the compressor to handle at a low duty cycle. Then too, you buy refrigeration with the intention of cooling things off. It doesn't take much food and beverage to completely overload a small system. Even in temperate climates duty cycle can be 50% and in the tropics, it isn't unusual to see compressors operating continuously ... until they fail, which is certain under a 100% cycle.

6.5.4 AC Refrigeration

Larger boats and RVs, or those with a big inverter will often use an AC motor to drive an open frame compressor. AC motors are less expensive than DC motors, and when tied to the dock, work directly from shore power. While an AC motor and open frame compressor is a good system, don't even think about an AC hermetic motor/compressor combination. The motors in hermetic sealed units tend to be of lower quality, unable to tolerate the voltage and frequency variations that are present with an AC generator, and to some extent, with the inverter. The hermetic motor/compressor is cooled by the flow of refrigerant. A large size condenser is required, and if it isn't properly cooled, then the motor burns up, mandating a costly repair. For best results, stick with the separate motor and compressor.

AC refrigerators normally use holdover plates since AC is not continuously available. When driven from an inverter, battery power may be used to pump down the plates. Generally, batteries will not be large enough to support refrigeration loads driven through the inefficiencies of AC motors and inverters. Thus, an engine is normally run whenever the inverter is powering the refrigerator.

Although an air cooled condenser may be used, the same efficiency issue arises as in the DC system. Water cooled condensers are always preferred, despite additional pump draw.

6.5.5 DC Compressor with Holdover Plates

The most energy efficient system is a larger DC motor and open frame compressor with holdover plates. A motor of 1/4 to 1/2 HP is sufficient. If the motor is matched to the compressor, and the motor/compressor to the holdover plates then a very efficient system results with minimum run time necessary. Water cooling is a must. The larger the DC motor, the more efficient will be the matching compressor, and the shorter the pumpdown times. Naturally, maximum current drain from the batteries will occur with a larger motor. The system may be run from batteries, or only when the engine alternator is charging. With enough wind and solar power, operation without the engine is possible.

A small problem arises with the large DC system when tied to shore power. To supply all of the motor current, a sizeable battery charger is required. If a small charger is used, it may be able to keep up with average draw, but when the motor is running, the batteries will be discharged. Such operation can result in quite considerable battery wear as the batteries are repeatedly discharged and charged. As you are probably aware, high output chargers that are suitable for full time hook up to the batteries are quite expensive.

A better solution is at hand. While connected to shore power, the DC refrigerator can be started from a timer and stopped as normal from a low pressure switch. An inexpensive unregulated charger can be operated at the same time as the refrigeration compressor, avoiding some or all of the battery discharge.

6.5.6 Insulation

The most important part of the refrigeration system is the insulation. You wouldn't know this by observing the amount of insulation provided by the typical boat or RV manufacturer. Unless you run so much food and beverage through the box that it barely cools, insulation losses will take more cooling power than actual foodstuff. Before you can expect to enjoy refrigeration, you must use plenty of insulation.

There is only one kind of insulation that is suitable ...polyurethane foam, 2 pound density. Other insulations do not have the same insulating qualities. Polyurethane foam is available in sheets, or in two parts liquid which react on mixture and can be poured into cavities. About 4 inches is suggested for refrigerator boxes while 6 inches or more is good for freezers. The book *Living on 12 Volts with Ample Power* describes how you can calculate loss due to insulation.

If you are designing a new box, do not use a front loading door. Besides loosing cold continuously through the door, every time you open the door, another significant loss of energy occurs as cold air falls out of the box. Even a top loading door looses some energy when opened so don't open it needlessly.

6.6 DC–AC Inverters

An inverter takes DC from the batteries and turns it into AC so that you can enjoy the use of everyday appliances without a generator. For the most part, inverters are efficient, from 85–95%. Still, most home appliances consume much power, so inverter use must be limited to short periods.

Inverters do not produce the same kind of AC that comes from the power company, or an onboard generator. Power from generators is si-

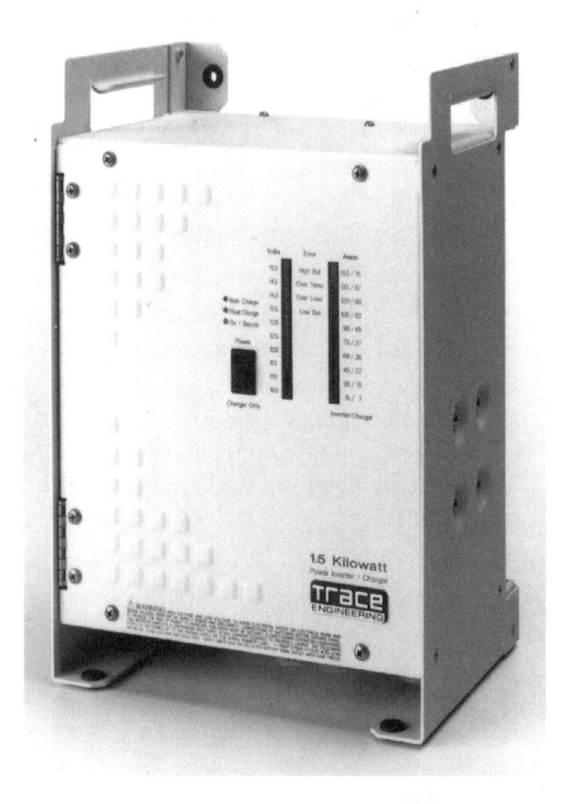

Figure 6.2: Trace 1500 Watt Inverter/Charger (70 Amps)

nusoidal, whereas power from most inverters is rectangular. Inverters produce a pulse of energy 120 times a second. Half of the pulses are positive, and half are negative. The width of the pulse is adjusted to maintain a given RMS voltage.

While most apparatus will work from pulses, energy is wasted due to harmonics. This wasted energy can also cause electrical noise that may interfere with sensitive radio and TV equipment. It may also penetrate audio equipment, showing up as a steady buzz. AC motors do not operate as efficiently from pulses as they do from sine waves, so expect more heating and shorter motor life. Sine wave inverters that produce utility grade power are available, but at a premium price. Sine wave inverters may also produce radio interference since the sine wave is derived by high frequency synthesis of the 60 Hz, low frequency wave. The synthesized 60 Hz wave is actually composed of hundreds or thousands of short duration pulses whose duty cycle averages to a sine wave.

Inverter specifications are not for continuous operation. Generally, an inverter will run at its specified power for about 15 minutes. After that, derate the unit about 25%. That is, for a unit that is labelled 1200 Watts, assume no more than about 900 Watts continuously. If you plan to power an AC refrigerator that will be running for longer than 15 minutes, be sure to derate the manufacturer's specification on the inverter. Since inverters don't like to drive AC motors, it is safer to derate the inverter specification to about 60% of normal, that is 720 Watts for a 1200 Watt inverter.

Not all inverters are as efficient as 85–90%. There are square wave inverters that can be about 50% efficient under heavy load. Even at idle, they may consume prodigious amounts of power. These inverters are priced much below a high quality unit. Nevertheless, an inexpensive unit can be used for very light usage if you are willing to turn it on before each use, and off immediately after use is complete. Small tools such as a drill or jigsaw can be operated from such an inverter. If you plan on frequent inverter use, or you want to have an inverter with low standby power that can be ready on demand, consider a high quality unit such as those made by Trace[2]. Their 1500 Watt unit is state–of–the–art and *best of class.* See Figure 6.2.

---

[2]Trace Engineering, Arlington, WA

6.7 Microwave Ovens

A microwave oven represents a huge load to the batteries. Usually, the oven won't be run for an extensive time, so total Amp hour consumption will be low. DC operated microwave ovens are available, although at a premium price. Since an inverter has other uses, such as operating a toaster, hair dryer or electric tools, if you plan on a microwave, you might as well get an AC one and an inverter.

Microwaves are specified in cooking Watts. To figure the amount of input power required for the oven, double the cooking Watts. That is, for 500 Watts of cooking power, plan on 1000 Watts input from the inverter. Assuming that the inverter is 90% efficient, actual Watts from input will be 1111 (1000/.9). That's about 92 Amps from a 12 Volt battery. To support 92 Amps for 15 minutes, a 400 Ah battery bank would be appropriate.

6.8 Toasters

Toasters make heat from electrical current ... always an inefficient operation. Toasters are available that operate from 12 Volts, but they take quite a while to do so. To run an AC toaster, you will need an inverter of at least 1200 Watt capacity.

6.9 Hair Dryers

Hair dryers are another device that come in 12 Volts, but they are also slow since they produce a minimal amount of heat. To run the bigger hair dryers from AC will require an inverter of about 1500 Watts.

6.10 Diesel Furnaces

Diesel furnaces are becoming more popular as people discover the advantages of forced air dry heat. The fans in diesel heaters represent quite a battery load, particularly on the larger heaters. Each time the furnace ignites, a glow plug is heated red hot. The glow plug can draw 20 Amps or more for about 30 seconds.

A diesel furnace can destroy itself if battery power is removed while combustion is occurring. The fan is absolutely necessary to remove heat from the exchanger. Because of the danger of power loss to the fan, furnace manufacturers suggest connecting the furnace directly to a battery. This means that a battery is more or less dedicated to the furnace, which is contrary to the practice of powering all loads though a main disconnect breaker or switch.

You can avoid a directly connected main battery by using a small third battery. A battery with only a few Amp hours of capacity is sufficient. The furnace gets tied directly to the third battery.

The third battery is connected to positive distribution with a high current, Schottky diode, so it gets charged whenever the house batteries are charged. If someone accidentally turns off the battery selector switch with the furnace running, the small battery will supply power for the fan long enough to cool the furnace. The furnace has a low voltage cutout so it will shut itself off when the small battery gets discharged. A schematic for this circuit is shown in the chapter on schematics.

6.11 Electronics–Radar

Most electronic equipment doesn't draw much power, although with enough gear, you can discharge batteries quite rapidly. The one item that does take quite a bit of power is the radar. Even on standby, the power producing tube in the radar is left on to avoid *thermal cycling* of the tube. Thermal cycling is more detrimental to the tube, than continuous operation. What you will save with the radar on standby is the power that it takes to spin the antenna, which may be 20% of the total draw.

6.12 Electric Windlass

When running, the electric windlass consumes much energy. Usually it isn't run for long, and most often, the engine is running at the same time so battery drain is minimal.

Occasionally we see batteries that are dedicated to windlass service. It never made sense to us for a number of reasons. Why carry around a lot of lead that is seldom used? Why load up the bow with lead ...if you need more weight up front, consider more anchor chain. Why put a battery in a place that guarantees acid slosh? Just what justification could there be for a windlass battery?

Recently the logic which circulates as folklore was explained by a caller who was convinced that a windlass battery was justified. By putting a battery in the bow, he could avoid running large size wires from the house banks forward. In fact he proposed to use #10 gauge wire. The battery itself would cost less than the large wire that would be required otherwise.

Well, if your alternator is puny, say 30 Amps maximum, then #10 wire to a windlass battery would work. But wire to a windlass battery must be sized for the maximum rating of the alternator! If the windlass battery gets discharged, which it most certainly will at some time or another, then it could receive all the alternator output. With a 100 Amp alternator, #2 wire or larger is required. That's probably big enough to drive the windlass directly, so why bother with the battery?

6.13 Entertainment Devices

If you are interested in power conservation, don't use normal household entertainment devices and run them from an inverter. Units which have been designed to operate from either AC or DC are more efficient than straight AC units. Automobile stereo equipment will be more efficient than home systems.

Power consumed by TVs is proportional to the size of the picture tube, so go with the smaller unit.

6.14 Summary

Loads are the reason we want electricity, but to avoid being a slave to our electrical habit, we must choose loads carefully, and manage them wisely. Special attention must be given to refrigeration since it may well be the biggest electrical load in the system. Fluorescent lighting offers the most light for power input, but costs more initially, and may be failure prone or electrically noisy. AC appliances are convenient, but they are inefficient and require a DC–AC inverter.

Chapter 7

Batteries

7.1 General Information

For the most part, batteries are taken for granted. In our automobiles, battery failures are only an occasional problem ... why should it be any different in our boats? Boat and RV designers, both amateur and professionals, don't usually consider batteries until it is too late to avoid serious compromises. Somewhere in the late stage of design, the designer poses the question, *where can I fit a couple of small batteries?* Convenient space is then sought, and the decision is made to use a given battery size. This is clearly a case of the tail wagging the dog. Rather than put batteries in the space available, the correct queston to ask would be, *how much battery capacity is required for the expected loads plus a reasonable margin for additions?* After the needed capacity is determined, then space must be made available or the user will suffer.

Whereas the designer leaves too little space for batteries, the user is quick to add loads beyond a reasonable hope of powering them except by outright destruction. Poor charging methods compound the problem of overloads. If the builder supplied any instrumentation, it is generally next to worthless. In the end, the user pays, over and over again as new batteries are purchased and murdered.

Too little battery capacity will always lead to premature failure, but knowledge of battery characteristics goes a long way towards minimizing problems. While we attempt to provide salient information in this chapter, the book *Living on 12 Volts with Ample Power* should be referenced

Temperature F/C	Absorption Voltage
122/50	13.80
104/40	13.98
86/30	14.19
77/25	14.34
68/20	14.49
50/10	14.82
32/0	15.24
14/(-10)	15.90
(-4)/(-20)	17.82

Table 7.1: Temperatue Compensation Table

for more detailed information.

7.2 Battery Temperature

Batteries like the same temperature as humans do. They perform poorly in cold weather, and are easily overcharged at higher temperatures. A battery should never be operated over 120°F. At this temperature vigorous gassing will result at 13.8 Volts, which is the voltage that many chargers are set at. This gassing will boil the electrolyte from the battery. The battery may even go into thermal runaway with the potential of explosion.

Voltages used in this chapter are correct at 77°F. Temperature compensation is required for other temperatures. Table 7.1 should be consulted for compensation amount. The table shows absorption voltage for 12–volt systems. Float voltage should be about 0.6 Volts less than the value in the table. For 24–volt systems, double the numbers given.

7.3 Specific Gravity

We are most familiar with liquid electrolyte lead acid batteries. Until recently, lead acid batteries came with fill caps, and the addition of water was a periodic chore. The chemical solution in a lead acid battery is a dilute mixture of sulphuric acid and water. The weight of the solution, or electrolyte, relative to water is called its specific gravity. For most fully charged deep cycle batteries, the electrolyte weighs about 1.265 times water, by equal volumes. The decimal point is omitted, and the specific gravity is stated as 1265 points.

As the battery is discharged, the specific gravity declines. Acid is absorbed into the plates, leaving a weaker solution externally. It is possible then, to measure the state of charge of a battery by measuring its specific gravity. More information about this subject is presented later.

Batteries that exhibit a great difference in specific gravity from cell to cell may need to be *equalized*, or replaced. The subject of equalization is covered later.

7.4 Deep Cycle Batteries

More than likely ·you are familiar with the term *deep cycle*. The description is given batteries that are designed to be deeply discharged in a cyclic application. For instance, an electric forklift draws heavily on batteries, and does so every day, thus the battery is cycled through deep discharges, followed by charges. The automotive battery is not designed to be deeply discharged ... in fact, a single deep discharge can ruin a starting battery.

To be called deep cycle, a battery should be constructed with thicker plates and the plates should be separated by high quality insulators whihc have enough porosity to readily allow electrolyte diffusion. By contrast, the starting battery is made with thin plates and paper separators, using manufacturing processes that are the least expensive.

Not all deep cycle batteries live up to the label on their case. Many manufacturers employ the materials and processes of the starting battery, with only slight differences. The quality of a deep cycle battery is generally reflected in its price.

7.5 Cycle Life

Some manufacturers make claims regarding the number of times that the battery can be discharged. You can be certain that you will not get the number of cycles they claim. Besides the inevitable inflation of data within marketing departments, you won't treat batteries to the ideal conditions that the manufacturer does when life tests are made.

How many cycles you obtain from a battery depends on the quality of the battery, as well as the treatment you give it. If you use batteries without instrumentation that informs you when to charge, and when to stop charging, expect short battery life. We've heard more than a few owners moan, "but these batteries are only six months old."

Cycle life is dependent on the depth of discharge. That is, you will get more cycles if you discharge only 30% than you will if you discharge 80%. This doesn't mean that you should only discharge 30%. At a 50% discharge, you will get more total energy from the battery over its life than at any other discharge. It pays to know how far discharged your battery is, so that you can recharge at about 50%. More about determining depth of discharge is presented later.

7.6 Capacity

Capacity is a measurement of how long a battery will supply a specified current. Current is measured in Amps, and time is naturally measured in minutes or hours. It used to be that battery capacity was specified at only 20 hours, that is, how much current can be drawn from a battery if it is to last for 20 hours. A 100 Amp hour (Ah) battery would supply 5 Amps for 20 hours. Some manufacturers still give the 20 hour rate.

Automobile batteries are specified in *reserve minutes*, generally at 25 Amps. That's because 25 Amps is considered a normal load on a stalled car that may have headlights on, and perhaps a fan for the heater. The reserve minute specification may not be a good indication of how long a battery will stand up under a deep discharge, particularly for a small capacity battery. If the manufacturer gives the length of time that a battery will operate with various loads, then it is easy to determine how long a discharge you can expect at any other current not listed. See the instrumentation section about Peukert's equation.

In general, battery capacity ratings are not to be trusted. Until recently, the consumer had no convenient way of testing Ah capacity, so battery manufacturers engaged in creative specifications. We've seen batteries specified that claim more Ah per pound of battery than state–of–

the–art processes allow.

7.6.1 Capacity Testing

Rather than rely on a manufactureer's specification, perform your own capacity test. With an Energy Monitor Controller, manufactured by Ample Technology[1], capacity testing is easy. If you don't have a meter with true Amp hour capabilities, then the task is more difficult, but worth the effort.

To perform a capacity test, you will start with fully charged batteries. Then a load is applied and the time measured before the battery voltage reaches 10.5 Volts. The average Amps, times the length of time of the discharge is the Ah obtained at the specific current. If you perform two tests, and aren't afraid to use a scientific calculator, you can use Peukert's equation to determine capacity at any discharge current. Peukert's equation is presented in the chapter about instrumentation.

Capacity testing should be done at least annually to determine how fast your batteries are aging. Once capacity falls to about 80% of new, think about replacement since batteries deteriorate rapidly past this point.

7.7 Discharging

As mentioned, you will get more total energy from a battery in its lifetime if you discharge it by 50%. This begs the question, how can 50% discharge be determined? There are three ways; by measurement of specific gravity, by voltage measurement, and by Ah measurement. Specific gravity and voltage are related by an arithmetic formula, so they amount to the same basic measurement. Direct Ah measurement requires computing power that can make current measurements at a high rate and perform exponential arithmetic needed to determine capacity remaining.

Neither voltage nor specific gravity measurements are accurate unless the battery has been rested for 24 hours. By rested, we mean, neither charged nor discharged. You just can't read the voltage of a battery as it is being discharged, and determine that it is at a specified depth of discharge. So how can voltage measurement be applied?

[1]Ample Technology, Seattle, WA.

By using two batteries, and alternating between them on a daily basis, you will have a rested battery which allows you to determine its depth of discharge. Before loading the rested battery, determine its depth of discharge. If it is about 50%, then it is time to charge. There is no precise voltage that indicates 50% on all batteries, but a voltage of 12.2–12.3 can be used as a rule of thumb.

With direct Ah computation, you can avoid the rested battery hassle. It turns out that there are many good reasons why a single house bank with a dedicated starter battery is the preferred system.

As long as you are using an accurate instrument, charging should normally take place at 50% depth of dsischarge. If you plan to motor somewhere and batteries aren't discharged to 50% you still want to charge them. You're not paying to run the engine just for charging, so top the batteries off any time the engine runs.

Lead acid batteries do not have the memory effect that some NiCad batteries exhibit. Discharging them to 50% each time will not result in the battery losing half of its capacity.

7.8 Charging

Charging batteries is not nearly as simple as it appears. Deep cycle batteries can not be fully charged and thereafter maintained by a single voltage. A voltage high enough to charge batteries fully, will soon destroy them after they are charged. To fully charge a deep cycle battery, an absorption voltage of 14.4 Volts is necessary. After charging, a voltage greater than 13.8 results in overcharge. Even 13.8 is too high if constantly applied, such as the case for a dockside battery charger. Depending on the charger technology, and the pattern of usage, a fully charged battery should be *floated* at 13.2–13.7 Volts.

It takes both voltage and current measurement to determine a full charge. A battery is fully charged when its voltage has risen to 14.4 and the current through the battery has declined to about 2% of its Ah rating. When voltage climbs to the absorption setpoint of 14.4 Volts, current through the battery will begin to decline quite rapidly. After 30–45 minutes, current through the battery will be low enough so that it is not wise to continue running a generator or alternator to continue charging. Although current will still be higher that 2% of its Ah rating, the economics of running an engine to top off the battery must be considered.

7.9 Equalization

Liquid electrolyte batteries that have been floated at a low voltage for long periods need to be periodically equalized. Equalization is the process that equalizes the specific gravity in all the cells. Basically, equalization amounts to a controlled overcharge.

The easiest way to equalize batteries is with a small unregulated battery charger such as those found at automotive parts stores. For 200 Ah batteries such as an 8D, a 10 Amp unit works well. Attach the charger to a fully charged battery and leave it on overnight. Before removing the charger the next day, observe the voltage across the battery. If it is 15.5–16.2 Volts, the battery should be equalized. A check of specific gravity will tell.

Since the voltage will rise above 15 Volts, it is good practice to only equalize an unloaded battery, as some equipment may be damaged by too high a voltage.

7.10 Liquid Electrolyte Batteries

Until recently, a lead acid battery was assumed to contain liquid electrolyte. In such a unit, the plates are suspended in a bath of electrolyte. To make the plates strong enough, particularly in a mobile application, the plates need to be thick. Antimony is usually added to the lead grids to harden the metal alloy. Antimony is a detriment ... it leads to a higher rate of self–discharge, and it lowers the charging voltage at which the battery begins to *boil*. Boiling is the term given to the evolution of gas that occurs whenever a battery is charged. The gas is hydrogen and oxygen, so liquid batteries should always have plenty of ventilation.

As a liquid battery sits unattended, it will slowly discharge due to the dissimilar metals present. This self-discharge is particularly damaging to a battery if it goes on beyond 30 days. Hard sulfate is formed on the plates, which can not be dissolved on recharge.

Liquid batteries have thick plates to provide the strength needed for deep discharges. The plate thickness limits the rate at which they can be discharged and charged. When used for engine starting, deep cycle batteries need to be oversized. During charge, electrolyte is slow to diffuse throughout the plate, so the rate of charge acceptance is lower than in thinner plate batteries. A longer, slower charge is required.

Liquid batteries can be damaged by 100% discharges, especially if subject to motion. The active material is solvent in the electrolyte when deeply discharged, so it tends to wash out of the plate grids and collect on the bottom of the battery. If enough washes out, then the plates short.

7.11 Sealed Batteries

So–called sealed batteries were introduced to the automotive arena a number of years past. These batteries are conventional liquid electrolyte batteries, but with enough liquid present to last the life of the plates. Such batteries are not appropriate for deep cycle application, and should not be confused with the sealed batteries that re–absorb oxygen within the battery. The latter sealed batteries are truly sealed and operate at a slight internal pressure.

When a battery is charged, the positive plate evolves oxygen, and if charged long enough, the negative plate evolves hydrogen. Sealed batteries use lead grids with a small amount of calcium as a stiffener. Unlike antimony that causes hydrogen generation before the negative plate is fully charged, the effect of calcium can be ignored. The negative plate in a sealed battery has a little more active material than needed, so under normal circumstances, the negative plate is not charged to the point where hydrogen is produced.

Oxygen is produced at the positive plate. Even in a liquid battery, some oxygen migrates to the negative plate where it is absorbed, but the liquid does not permit sufficient oxygen migration to eliminate oxygen loss into the environment. In the sealed battery, electrolyte is porous, allowing oxygen to migrate in gaseous form to the negative plate. With enough porosity, all the oxygen can reach the negative plate and be absorbed. As long as there is no hydrogen generated, no gases escape the battery.

Electrolyte porosity is attained by at least two methods. Original patents were issued to Sonnenschein[2] of West Germany in 1933 for a

[2]Sonnenschein, Büdingen, Federal Republic of Germany.

gelled electrolyte. The gelled electrolyte forms fissures which allow direct oxygen passage between the plates. Another technique for providing oxygen passages involves the use of a fibrous separator that is saturated to about 90% with electrolyte. Open channels along the fibers allow oxygen to reach the negative plate.

Sealed batteries have many advantages over liquid ones. Liquid batteries that are floated for long periods of time at a low voltage need to be *equalized*, a process of applying a high voltage to the battery so that the specific gravity reaches the same value in all the cells. Equalization is a hassle to do correctly, and can be harmful as normally done. Sealed batteries do not need to be equalized.

Sealed batteries have a higher rate of absorption than do liquid batteries. That is, you need to put back in less Amps for a given discharge. That means that sealed batteries charge faster using the same charger. Sealed batteries also accept a higher rate of charge ... you can use a bigger charger without fear of overheating the sealed battery.

Naturally, a sealed battery will not leak acid in your boat. Acid won't eat holes in your jeans when you work around the sealed battery either, and you won't have to put up with the rotten egg smell of sulphur during recharge. Liquid batteries, when charging, produce a couple of toxic gasses, stabine and arsine, as well.

Sealed batteries will discharge, but won't be ruined if submerged in saltwater, and running them flat and leaving them that way for 30 days won't mean you'll have to buy new batteries. A liquid battery that stays completely discharged for more than a few days is ruined.

Sealed batteries also stand up better to heavy currents and deeper discharges than the liquid battery. In practice, sealed batteries appear to live as long, or longer, although some manufacturers have taken to quoting a great number of discharge cycles for their batteries.

With all these advantages, what is the downside for sealed batteries? Would you believe cost? While initial expense is high, cost per Amp hour over the life of the battery is equal or less than that of the liquid battery. Whether you select sealed batteries, or elect to use a high quality liquid unit, protect your investment with proper instrumentation and charging equipment.

7.12 Golf Cart Batteries

Golf cart batteries are widely used for marine and RV applications. But are they deep cycle? Not really. For the most part a golf cart battery is a close relative of the starting and lighting (SLI) battery. Manufacturing cost is a prime factor. That means SLI practices such as thin cast radial grids, minimum separator resistance, and heat sealed plastic cases are used. The object is to obtain high energy density for maximum discharge rates. In the process, deep cycling capability and battery life are sacrificed.

Golf cart batteries are also of the liquid electrolyte variety. Despite their relatively short life, however, golf cart batteries are suitable for some systems. If you can't afford the sealed gelled batteries, golf cart batteries may be the alternative you are looking for.

7.13 Heavy Duty 8D Batteries

Heavy duty 8D batteries are a close relative of the golf cart battery. Our first batteries were two heavy duty 8D units. We got six years of service from them, including two years of cruising when no dockside charger was plugged in. We got this great service by monitoring their condition accurately, and charging when needed to a full state. Many users are happy to get two years of service from batteries ...proper charging and good instrumentation does pay off.

7.14 Sealed Battery Experience

Some of the first cruisers left in 1987 with sealed gel batteries. We recently inspected a set of those batteries and for all practical purposes they performed as new in 1995. They were given exceptional care, however, with temperature compensated charges and a watchful eye on state–of–charge during discharges.

When a sealed battery is overcharged, it will emit some gasses. Done long enough, the battery will dry out. You can't add electrolyte, so the first secret to longevity is to avoid overcharges. This means that the voltage must be accurate for the battery temperature.

Deep discharges are not as detrimental on some gel batteries as they are on liquid units, but monitoring the Amp–hours remaining is still important to get long life.

No other lead–acid battery will accept a fast charge like some gel units. Even so, the gel battery will sometimes refuse to take a charge after long periods of inactivity. The cure for this is to place a small load on the battery and discharge it completely ... until the voltmeter shows less than 5 Volts. Now recharge the battery slowly and until it reaches a full charge. Stubborn batteries may take 2–3 such discharge/charge cycles to recover normal operation. Voltage should hold at about 13 Volts overnight or longer if the battery is to yield full capacity.

After doing the above treatment and the gel battery still refuses to hold about 13 Volts, all is still not lost. Turn the battery bottomside up and repeat the discharge/charge cycles. We've done this on several occasions and recovered batteries that would have been scrapped otherwise. After two years, we're stilling using a 4D that was recovered only after a treatment bottomside up.

7.15 Watt–Hours versus Amp–Hours

Most of us like to think in terms of the Amp–hours that a particular battery bank will provide. Even this is made difficult by the battery manufacturers who continue to rate deep cycle batteries using the same scale that is used for starting batteries ... reserve minutes.

As users of deep cycle batteries, we're more interested in how long a battery will power loads over an extended period of time, say from 20 to 100 hours. It is customary to rate traction batteries, those batteries used for forklifts and other material handling equipment, in terms of Amp–hours. Amp–hours is just the simple product of Amps multiplied by time. Why Amp–hours? Aren't we really interested in the amount of work that a battery will do before it needs to be charged?

For electrical devices, work is measured in Watt–Hours. Watt–Hours, or Wh, is the product of Amps times Volts multiplied by time. This is obviously a more complex calculation, and one that would have been difficult to do with just the mechanical devices of 75 years ago when battery research was in full stride. Early Amp–hour meters were constructed using the single disk Faraday motor shown earlier in the book. Permanent magnets were used to create a magnetic field that cut the disk. With DC motors, rotational velocity is proportional to the amount of current flow. The Faraday motor has a single turn winding, the disc, and so it exhibits a very low resistance, more or less like a shunt. Once calibrated, it's a simple matter to count the revolutions and scale them to Amp–Hours. The motor meters were designed so that they ran slower in one direction. By adjusting this speed, researchers could determine approximate battery Ah efficiency under standard laboratory conditions.

With the modern microcomputer at our disposal, rating batteries in Watt–Hours instead of Amp–Hours is easily accomplished, but why would we want to? A Wh rating is a better indication of performance since it is a direct measurement of the *work* capacity of a battery. As mentioned above, Wh includes battery voltage as a parameter. Battery voltage is important for just about every electrical device that operates. Inverters are more efficient at higher battery voltage. Other devices are also more efficient. Take the lowly water pump. Assume that it has an equivalent resistance of 1.2 Ohms so that it draws 10 Amps at exactly 12 Volts. At 11 Volts the pump will only draw 9.167 Amps.

The amount of water that can be pumped is proportional to the amount of energy that the pump consumes. Suppose that we have a quantity of water that can be pumped in 1 hour at 12 Volts and 10 Amps. That would consume 10 Amp–hours and 120 Watt–hours. While the pump's efficiency may actually worsen at a lower voltage, we can ignore that and still demonstate our point about Watt–hours and Amp–hours. Just remember that the actual result will probably be somewhat worse than calculated. To pump the same amount of water at 11 Volts, we'll still have to provide 120 Wh of energy since work is equivalent to Wh. At 11 Volts and 9.167 Amps, the pump will have to run for 1.19 hours. Now multiply 1.19 times 9.167 and you find that the pump consumed 10.9 Amp–hours instead of 10. With a lower voltage, the pump not only runs longer but consumes more total Ah for the same Wh of work! This means that a battery which holds its voltage higher throughout the discharge cycle will have more *apparent* Amp–hours.

Characteristic	Thick Plate (Traction Type)	Medium Plate (Golf Cart) (Heavy Duty)	Gel Type (Prevailer) (Gel One)
High Rate Discharge	Poor	Good	Excellent
High Rate Charge	Poor	Fair	Excellent
Charge Efficiency	Poor	Good	Excellent
Watt–Hour Efficiency	Poor	Good	Excellent
Capacity vs Rate of Discharge	Poor	Good	Excellent
Maintenance	High	High	Low (none)
Self–Discharge	High	Medium	Low
Deep Discharge Tolerance	Poor	Poor	Excellent
Over Charge Tolerance	Fair	Fair	Poor
Under Charge Tolerance	Poor	Poor	Good
Cycle Life (Marine Use)	Low/Medium	Low/Medium	High
Cost per Amp–Hour	High	Low/Medium	High
Safety (Gassing)	Low	Low	High
Paralleling Units	Good	Poor/Fair	Excellent
Float Voltage	13.6 Volts	13.6 Volts	13.5 Volts
Gassing Voltage	13.9 Volts	13.8 Volts	13.8 Volts
Absorption Voltage	14.4 Volts	14.4 Volts	14.2 Volts
Equalization Voltage	16.2 Volts	16.2 Volts	N/A
Rested 'Full' Voltage	12.6+ Volts	12.6+ Volts	12.9+ Volts

Table 7.2: Battery Comparison Guide

7.16 Battery Comparison

How can we compare battery types? Because there are so many different manufacturers with different processes, it becomes a difficult task. Table 7.2 is an attempt to compare some features and it also gives some useful voltages for charging.

Thick plate batteries, sometimes called traction batteries, will generally give longer cycle lift than thinner plate units. This generality falls apart when thick plate batteries are used at either high rates of charge or discharge. Traction batteries would like to be charged at about 10% of their Ah rating, and discharged at not more than about 15%. Batteries in marine use are typically charged at high rates and cycle life suffers as a result. For solar powered systems, charge rate is generally slower and thick plate batteries are a good choice for extended life.

7.17 How Much Capacity?

We use two rules of thumb to "ballpark" how much capacity is needed.

- Amp–hour capacity should be four times daily consumption; and
- Amp–hour capacity should be four times the highest current load.

If daily consumption is 50 Ah, but you have a load that draws 100 Amps, then 400 Ah capacity is appropriate. Here is a good case for a gel unit since they will support high current loads more readily. With a gel battery, 200–300 Ah would be appropriate.

7.18 Summary

Professional and amateur boat and RV builders generally decide on battery capacity according to available space after all the rest of the boat's equipment is placed. Battery capacity must be considered based on need, not available space.

Specific gravity is related to the state of charge on a battery, as well as the general health.

True deep cycle batteries are constructed with better and stronger materials than the engine starting battery. Not all batteries with deep cycle labels live up to the label.

Starting batteries may not be deeply discharged more than a few times. Deep cycle batteries have a limited number of deep discharge cycles, ranging from 100 to 1000, depending on their quality.

Battery capacity is a measure of how many hours a battery will supply a specified current. Capacity at 20 hours is acceptable, but capacity for longer and shorter periods is useful. It is best to perform your own capacity tests.

Batteries are best discharged to 50%, for maximum energy over the life of the battery. The 50% discharge point can be determined by resting batteries, or by direct Amp hour measurement.

Charging batteries is more complicated than hooking them to a battery charger with a single, fixed, output voltage. Multi–step charging is required.

The disadvantages of liquid batteries, such as the periodic need for equalization, and slow recharging can be avoided by using sealed batteries that absorb oxygen internally.

Golf cart and heavy duty 8D batteries can be used, but long life is not likely unless good instrumentation is used to monitor discharges and charges.

A few tips regarding gel batteries has been presented, and the difference between Watt–hours and Amp–hours has shown that Watt–hours is a better measurement of battery capacity.

Finally, a comparison chart has been presented that rates battery types according to different characteristics.

Chapter 8

Charge Sources

8.1 General Information

Charge sources are any devices that supply a charge or source of energy. Included are battery chargers, alternators, solar panels, wind/water generators, and even pedal powered generators. Each type of source is covered in greater detail in subsequent paragraphs.

8.2 Battery Chargers

8.2.1 Charger Types

There are several types of battery chargers available. They may be classified according to the type of regulation that is employed. The simplest charger is nothing more than a transformer and a set of rectifier diodes that convert AC to pulses of DC. These chargers, found in automotive parts stores, are unregulated. The next level of sophistication is found in ferroresonant chargers that derive a poor quality regulation via a magnetic shunt on the transformer and a secondary that is tuned with a capacitor. Yet higher regulation is achieved with SCR phase controlled chargers. The highest level of regulation and performance is achieved by chargers that produce pure DC.

8.2.2 Ferroresonant Chargers

It may come as a suprise to you that the greatest number of battery chargers installed by boat manufacturers are the least qualified as battery chargers. These units are based on ferroresonant technology which is about 75 years old. As you know by now, charging batteries properly is more complex than hooking up to a constant voltage. If indeed the ferroresonant chargers produced a constant voltage, they wouldn't be so bad as they are, but the ferroresonant charger is a long way from a constant voltage device. The term *Finishing Voltage – 13.8 Volts* is a bad joke. At best, a ferroresonant charger is 5% regulated. That really means final output is anywhere from 13.1–14.5 Volts. On the one extreme, little charging is done. On the other, all the water will soon be boiled from the batteries with permanent damage the consequence.

Besides poor regulation, ferroresonant chargers don't produce much current at the voltages required to charge batteries. As a rule of thumb, if a ferroresonant charger has a label that declares it to be a 20 Amp charger, plan on about 1 Amp at 13.8 Volts ... a mere 5% of its rating. Of course the manufacturer will tell you that their chargers taper off as the battery gets charged as a safety measure. It's the old story ... talk up a failure as a feature.

Batteries will last for many years if treated properly. With a ferroresonant charger attached, the batteries are constantly cycled. Because the charger won't put out much current at charging voltages, each time you use an appliance, you are discharging the battery. At some point, you'll remove the load draw, and the charger will recharge the battery. Liveaboards cycle their batteries every evening. It may not be a deep discharge, but every discharge adds up.

If you are buying a new boat or RV that comes standard with a ferroresonant charger, you might begin to wonder about what other cheap thrills the manufacturer has built into the product. If you presently own a boat or RV with a ferroresonant charger, and find yourself adding water to the batteries more than twice a year, realize that you are paying the price of a good charger slowly, one battery at a time. Ferroresonant chargers should have disappeared with the advent of the Silicon Controlled Rectifier (SCR) back in the early 1960's. That they're still around 30 years later only demonstrates the lasting power of ignorance.

8.2.3 SCR Phase Control Chargers

The SCR is a device that rectifies like a diode, but must be triggered into the conduction state. It is used to regulate battery charging by control circuits that time the trigger synchronous with the AC frequency. The SCR is triggered into conduction just soon enough on each cycle to yield an average current that is sufficient to maintain a constant voltage on the battery. Because the control circuits are electronic, precise regulation can be obtained. In fact, regulation is about 100 times better than the ferroresonant charger. If the regulation is fixed at a single setpoint, the SCR charger is still not effective in delivering a fast full charge and thereafter protecting from overcharge. By now you know that multi–step chargers are required.

The SCR charger doesn't taper off like the ferroresonant unit, so you will get much closer to the label rating. Most SCR units will not produce full rating at low AC input, however, which is a common circumstance in many marinas and RV parks. On cold days with lots of heaters running that drag down the AC line, you may find your charger on, but your batteries flat.

SCR chargers do suffer a few other problems. Because they only conduct for a short time on each cycle, charge current is a series of high current pulses that the battery must average. This tends to heat up the battery, especially toward the end of charge. High current pulses also generate considerable radio and television interference, and will often cause a hum in stereo equipment.

Because an SCR can not be triggered off, phase controlled chargers are not generally short circuit proof. Some such chargers must be turned off before starting the engine, or you blow a fuse or breaker, and maybe the charger itself.

Note that inverter/chargers manufactured by Trace[1] and other popular brands use phase control techniques in their battery charger circuits.

8.2.4 High Frequency Switchmode Chargers

The highest level of performance is obtained by high frequency switching charger. These chargers rectify the AC line voltage and then switch it

[1] Trace Engineering, Arlington, WA

at high frequencies through a transformer to the battery side. Because of the high frequency, the transformer is much smaller than one which operates at 60 Hz.

A switchmode charger often produces pure DC ... the battery is not required to average high current pulses. Battery heating is therefore not such a problem. With pure DC across the battery, noise in stereo and other equipment is not present. A pure DC charger is also better when the batteries are being used at the time of charge, such as in a liveaboard situation. In such circumstances, the charger supplies the full load current and doesn't require the batteries to average charge pulses and load draw in a continuous cycle of battery wear.

Presently, switchmode chargers are based on high frequency techniques that were made popular in computer power supplies. The high frequencies may cause some radio or TV interference. Often, switchmode units are packaged in a minimum space that requires fan cooling. The fan may generate audible noise, so some consideration must be made for mounting the charger outside of living quarters.

Switchmode chargers are short circuit proof, so no precautions need be taken when starting an engine, or placing another heavy load draw on the battery. Some switchmode chargers are also over temperature protected, and some will produce full output current even with very low AC line voltages.

Figure 8.1 shows a multi–step, switchmode charger manufacturer by Charles Industries[2]. Note the generous heatsink which covers both sides of the unit. Because of the heatsink, no fan is required. Although rated at 20 Amps, we've seen them produce as much 23 Amps. The unit maintains full output even with low AC input voltage and is not sensitive to the AC frequency.

8.2.5 Power Factor

Power factor is a phenomenon that increases the reactive power being consumed. All chargers suffer from poor power factor, that is they draw more *apparent* power from the AC input than might be expected from the output current they provide. This means that a small generator can not operate as big a charger as wattage labels indicate. Whereas a

[2]Charles Industries, Rolling Meadows, IL

Figure 8.1: Switchmode Charger from Charles Industries

charger might be 75% efficient, a poor power factor makes it appear to the generator as only about 55% efficient. This means that you need a larger generator than expected, or you must be satisfied with a smaller charger.

Electronic methods to improve power factor have just been introduced to the integrated circuit level. We hope to see them incorporated into battery chargers soon so that maximum efficiency can be made of small portable generators.

8.2.6 Charger Tech Tips

Some people get by with small unregulated chargers. They have learned not to leave the charger connected except during actual hours that they have loads turned on. Depending on the load draw, the batteries in this situation may be over or undercharged. Used intelligently, however, an unregulated charger can provide reasonable service. Don't go too long between charging ... never over 30 days. Before leaving, run the charger long enough to return a full charge.

Small unregulated chargers can be used as an inexpensive way to equalize liquid electrolyte batteries. A 6–10 Amp unit can be left on the batteries for 12–24 hours. Let the charger peak out at 15–16 Volts. Monitor the battery temperature, and cease charging if the battery feels warm to the touch.

Unregulated chargers are also great for charging from small portable generators. With the generator running, you won't have the charger on long enough to overcharge. The cost of unregulated chargers is minimal, compared to fully regulated units.

An unregulated charger can be turned on and off at programmable two state–of–charge setpoints using the Energy Monitor Controller made by Ample Technology[3]. This method of charge control is as good as full multi–step regulation, but it is effective in preventing overcharges. The cut–off setpoint is also temperature compensated.

Under no circumstances can we recommend a ferroresonant type charger. If you have one, and can't afford to replace it, treat it as if it were unregulated ... for all practical purposes, it is.

---

[3]Ample Technology, Seattle, WA

It's normal to add water to liquid batteries occasionally ...perhaps twice a year. If you never need to add water, you are not charging the batteries fully. Batteries can be murdered from undercharges as easily as overcharges.

8.2.7 Battery Charger Cost

You'll pay the price of a good battery charger, whether you buy one or not. If you don't buy a good charger, you'll end up getting poor service from the batteries, with repeated replacement.

You are familiar with the supermarket practice of quoting the price of a commodity per ounce, or per volume. If battery chargers were sold in supermarkets, we could expect that they might carry a label spelling out the *dollars per Amp*. As mentioned, the Amps declared on the label of most chargers is usually not the usable Amps at 13.8 Volts. If you compute cost of a charger on an Amps per dollar basis, you find that the lowest cost chargers actually cost as much as 5 times that of a good charger! No one we know would buy laundry soap that is priced 5 times the nearest competitor.

When you add in the daily wear and tear that cheap chargers put on batteries, their true cost can't readily be calculated. Pay the initial price of a good multi–step regulated charger, and save on battery costs and headaches.

8.3 Alternators

8.3.1 General Information

The conventional automotive and marine DC alternator is actually a three phase AC alternator with rectifying diodes to produce DC. The alternator has a three phase stator winding that generates the output current. Rotating inside the stator winding is an electromagnet. The spinning electromagnet is called the *rotor*, and it causes current to flow in the stator. The rotor contains a winding called the *field*. By varying the amount of current flowing in the field winding, the output of the alternator can be controlled. Current in the field winding is about 3–6% of the output current.

8.3.2 Automotive Alternators

Automotive alternators are equipped with a regulator that senses alternator output and adjusts field current to control the output. The regulator is normally mounted on or in the alternator case. When the case temperature rises, the regulator cuts back the alternator output to prevent high temperatures which can destroy an automotive alternator. Charging batteries quickly is not a consideration in the auto, because the battery is never deeply discharged. The regulator in the auto operates at a single constant voltage. That voltage is chosen so that if you drive less than 150 miles a week, your batteries will be undercharged. If you drive more, then you will overcharge the batteries. A single setpoint regulator costs very little to make ... a key feature of any automotive part.

Alternator repair organizations often sell alternators that are rewound to achieve a higher rating in a standard case size. The cases and bearings for these alternators were not designed to handle the extra heat. We once evaluated a 100 Amp alternator that was formerly a 60 Amp unit. It took all of 15 minutes before we had black smoke pouring out of the windings. If an alternator wasn't designed to handle large and continuous loads, don't waste your money.

8.3.3 Deep Cycle Alternators

What may work for the automobile is not very appropriate for batteries that are deeply discharged on a regular basis. First, a fast charge is definitely desirable, and a full charge is necessary ... without damaging overcharges. The alternator, with high output current must tolerate higher operating temperatures, and still provide enough air flow to prevent meltdown. The regulator must provide a fast, full charge, and then switch to a lower float voltage to maintain a full charge, without overcharging.

Until recently, automotive alternators were used exclusively in the RV and marine industry. In some cases, automotive alternators had changes to their brush assembly to prevent sparks, and were called marine alternators. Even today, most boat manufacturers use engines that come standard with the typical automotive alternator.

Since high output alternators are designed to operate at elevated temperatures, the regulator is usually supplied as a separate box. Many marine alternators come with a single setpoint regulator. To obtain a full charge, the regulator setpoint is about 14.4 Volts. The regulator is still too inefficient to provide a fast charge, but you will eventually get a full charge. If you run your engine for long hours, you'll also destroy the batteries from overcharge.

For best results, a multi–setpoint regulator is necessary. The regulator should apply a bulk charge, an absorption cycle, and then trip to a lower float voltage. In short, the regulator should be right for batteries, not necessarily cheap to make.

8.3.4 Alternator Tech Tips

High output alternators need more belt tension than puny automotive units. Always use a belt rated as heavy duty. Such a belt will have a cog like backside which allows better belt cooling, and pulls more power around small pulleys. The Gates[4] heavy duty *greenstripe* belts have proven dependable.

If possible, use a 'B' belt, which is about 1/2 inch wide. An 'A' belt is slightly narrower at about 7/16 inch. Most pulleys today are cut to take either an A or B belt. If the wider belt doesn't seat into the pulley enough to provide full edge contact with the pulley, stick with the A belt.

Always use dual belts if possible. Belt tension can be less, so maintenance intervals are shorter. Belts are sold in matched sets ... shop elsewhere if a clerk tries to sell you two belts supposedly the same length. Make sure the package indicates a matched set.

High quality alternators can rotate at about 10,000 RPM under load. It is tempting to put a small pulley on the alternator and a big one on the engine to get full alternator output at idle. A pulley diameter less than 2.5 inches on the alternator can cause problems with belt slippage. If you keep the belt tight enough to avoid slip, then you are creating undue side load on the water pump and engine crankshaft.

Some people add high output alternators in front of the main engine, driving them from an auxiliary pulley on the engine. Such an alternator should be mounted to the engine chassis so that it vibrates with the engine. If not, bearing life in both the engine and alternator can suffer,

[4]Gates Rubber Company, Denver, Colorado.

and belt wear will be greater. Make sure you observe the direction of rotation ... more information about fan rotation is given below.

An alternator should be sized according to the capacity of the batteries that are being charged. As a rule of thumb, use an alternator that is 25–40% of the Ah rating of your house batteries. That is, for 400 Ah of capacity, use an alternator that is 100–160 Amps. If you are using sealed gel batteries, then the alternator can be 50–100% of the Ah rating.

Alternators are not very efficient, partly due to the power factor issue. When driven by small engines, a small frame alternator is about 30% efficient. A large frame alternator will be about 45% efficient. To compute the Amps which a small engine can produce, multiply its horsepower by 746 to obtain Watts. Divide the Watts by 14.4 Volts, which is the full charge voltage for 12–Volt batteries. Now multiply by the alternator efficiency. For instance, 5 horsepower is equivalent to 3730 Watts. Divided by 14.4, the current is 259 Amps. For a small frame alternator, only 30% of that is attainable, or about 78 Amps.

When viewed from the front, fans on alternators are slanted to move maximum air when rotated clockwise. With dual engines, one engine rotates counterclockwise. The standard alternator fan is not efficient if rotated backwards, and the high output alternator will burn up. You can get fans that have no angle on the fan blades, and are therefore equally efficient in either direction. They aren't as efficient as a slanted fan, however.

Dual output alternators sacrifice air flow through the alternator. Failures due to heat are much greater. An isolator is a preferred method to charge multiple batteries from a single alternator. A regulator that compensates for the isolator is necessary.

It is not good practice to run high currents through engine brackets and blocks. To avoid this, connect a good ground wire to the alternator case. The ground wire needs to be as large as the output lead. Some alternators are available with isolated grounds, so the ground wire is required. It is best to run the ground wire back to negative distribution.

Be sure to use nylon locking nuts on alternator output posts. Route wires to the alternator so that wire vibration will not loosen the nuts. A loose nut allows the wires to start arcing, and can quickly sever the post, just like an arc welder. If the loose wire happens to be the postive wire, and it falls against the alternator case, standby for buckled batteries and

Alternator Amps	Wire Size (AWG)
50	#6
90	#4
130	#2
160	#0
200	#00

Table 8.1: Alternator Wire Size

perhaps a fire. A Fail Safe Diode, or isolator can prevent battery discharge in such a circumstance. Large wire sizes are especially prone to vibrate loose on the alternator. A short piece of welding cable is quite flexible, and can be used to connect the alternator to the alternator shunt or other termination point.

The nuts which hold pulleys on, are fine threaded. Pulleys are not normally keyed anymore, so the nuts must be very secure to avoid pulley slip. If you have occasion to change a pulley, use an impact driver and socket. They generate much less stress on the fine threads, and don't tend to twist the rotor shaft like a manual socket wrench does.

Alternator wire sizes for the main output are way too small from engine manufacturers. When you add a high output alternator, be sure to change the wires, both positive and negative. Table 8.1 shows appropriate wire gauges for various high output alternators. For wire runs greater than 15 feet, consult wire tables in this book, or *Living on 12 Volts with Ample Power.*

Alternators can produce more current when cold than hot. Always use the cold rating to determine wire size and isolator ratings.

8.4 Wind/Tow Generators

Wind and tow generators use a DC motor. When torque is applied to a permanent magnet DC motor, the motor becomes a DC generator. The DC motor can be driven by either a wind or water propeller. Two bladed wind props are most efficient, but three bladed props are less noisy.

Wind/tow generators can produce a significant amount of power . . . enough

Figure 8.2: Ampair Wind Generator

to destroy batteries. They are also difficult to regulate. Their output can't be opened by a series regulator, because the prop will race and self–destruct. The output can't be shorted by a shunt regulator, because the permanent magnet may be demagnetized, destroying the motor. Most people regulate the wind/tow generator by monitoring the batteries and removing the generator when fully charged.

Most wind generators can not tolerate wind speeds much greater than 25 knots. Props disintegrate, and motors burn up at higher velocities. Some wind generators[5] use a centrifugal clutch like those in go–carts. When the wind reaches a critical velocity, the clutch imposes a drag on the prop to prevent runaway. Such a mechanism is not precise, so some performance is sacrificed on the side of safety.

Other wind generators, such as the Ampair[6] unit shown in Figure 8.2 are designed for low output and tolerance for high winds. These units can be permanently mounted on mizzen masts, or stern posts. While these generators don't produce as much power, they are much less hassle to use since you don't need to worry about wind velocity. They can still overcharge batteries, but not as quickly, so a regulator is recommended.

A new wind generator from Southwest Wind Power[7] has been designed from the ground up to yield a highly functional generator without the

[5]Windbugger, Key Largo, Florida.

[6]Ampair - Contact Jack Rabbit Marine, Stamford, CT.

[7]Southwest Wind Power, Flagstaff, AZ.

Figure 8.3: Air Marine Wind Generator

problems of earlier designs. Refer to Figure 8.3. A new coil configuration with rare earth permanent magnets provide immense power in a compact size. The unit uses dynamic braking to limit RPM and can therefore be left unattended. Because of this braking, the unit can be rougly regulated and can have its output opened circuited without fear of the prop racing to destruction. This means that solar panel controllers that switch on and off can be used without diversion loads which were required in the past to keep the generator loaded. Be advised, however, that the Air Marine unit can produce over 40 Amps peak so a relay rated for 40–50 Amps is required for reliable operation.

8.5 Solar Panels

Solar panels produce DC current from sunshine. The more the sun, the greater the current. The panels need be aimed at the sun to achieve greatest output, so careful attention must be given to placement.

On a dollars per Watt basis, solar panels have steadily declined until recently. No price breakthroughs appear on the horizon, but even so, solar power is effective, and reasonably priced in small systems.

Solar panels do require regulators, despite the claims of at least one company[8] that they have a *self–regulating* panel. Self–regulation, it turns

[8]Arco Solar, Camarillo, California.

out, is nothing more than fewer cells in series, lowering the output voltage somewhat.

When purchasing solar panels, be aware that output declines with elevated temperature. Be sure that the panel will produce a full charge where you intend to use it.

Solar panels require a diode in series, or they will discharge the batteries when the sun doesn't shine. Some panels come with diodes, while others don't. If you have panels without diodes, select a Schottky type diode for maximum efficiency. Be sure that it is generously overrated for the maximum current that the panel can produce.

Solar panels are made using a rigid crystalline cell, and a flexible amorphous thin film cell. Both types of cells lose efficiency as they age, but the amorphous types appear to suffer much more, losing as much as 30% in the first year of operation. When you consider that flexible cells cost almost twice as much per Watt, and use about twice as much surface area for the same output current, the advantage of flexibility is costly. The amorphous cells are, however, less affected by partial shading, so in some circumstances, the additional cost is justified. Some reports indicate that thin film solar cell lose as much as 30% of their output in the first year, so be sure to budget for enough current after degrading.

8.6 Pedal Generators

Despite human conceit, the power that a person can produce is quite limited. Bicycle generators are available, but don't expect to do much generation with one. If you have an ulterior motive to burn up a few calories, and produce enough power for the stereo or TV while doing it, then by all means, pedal.

Pedal generators use DC motors similiar to the wind/tow generator. With suitable gearing, you could adapt such a motor to a bicycle frame. Well trained athletes can produce about 1/4 horsepower or about 187 Watts. Pedal Generators are sold by Real Goods[9].

[9]Real Goods, Ukiah, California.

Chapter 9

Instrumentation

9.1 Introduction

In our book *Living on 12 Volts with Ample Power*, we touched on the need for instrumentation, and gave the voltages and currents that have significant meaning in a battery system. We also explained the importance of measuring current.

In this chapter, we explain the principles of instrumentation, why you should instrument, and we explain how to achieve a level of instrumentation that suits your system and style.

9.2 Why Instrumentation?

Columbus didn't have much instrumentation at his disposal, yet he managed to bring some of the first tourists to the Americas. Of course, he didn't have electricity to contend with either. If you intended to cruise in the style of Columbus, you wouldn't be reading this book. Electricity is assumed. Indeed, electricity is as essential to a modern lifestyle as medicine. Simply put, instrumentation allows us to manage electricity. With proper management, electricity will cost less ... no more murdered batteries, no more sudden failures.

Directly, we can't see electricity. In low voltages, we can't feel it either. Nor can we taste it, or smell it, unless it runs wild. To manage electricity, we have to sense it. We must use indirect methods that let us *see* electricity.

9.3 What do we Measure?

From earlier chapters, we have become familiar with the basic principles of electicity. We learned Ohm's law that applies to a *closed* circuit, and we learned that our loads represent resistance, one entity in Ohm's law. The other two entities are voltage and current. By knowing any two entities of Ohm's law, the third can be calculated.

Since our loads are resistance, and since we turn loads on and off, it makes sense to measure voltage and current. In practice, we only care about the resistance of our loads if the batteries are overloaded. We do care about the voltage and current in the system because their product is power, that is, by knowing both the voltage and current, we can determine how much power we are drawing. Since batteries contain a limited amount of power, we must know how much we are consuming.

Battery voltage has long been measured. Current, the other half of the picture, has largely been ignored. Does this imply that current is not significant? Not even. For the most part, partial information is better than no information. Voltage is much easier to measure than current, consequently, much less expensive. If economics dictate partial information, then voltage it is. With battery systems, however, leaving out measurement of current is false economy.

Voltage measurements are extremely useful in a battery system. A measurement of 14.4 Volts on the battery doesn't indicate that the battery is fully charged, but it does indicate that some source is indeed charging the battery, and given enough time the battery will reach a full charge. A measurement of 12 Volts on a battery may mean that it is heavily loaded, or it may mean that the battery is really discharged and it won't be long before collapse occurs.

Current measurements are equally meaningful. A positive current of 50 Amps through a battery indicates that the battery is being charged, and that it will eventually reach a full charge (if not already overcharging). Likewise, a current drain of 50 Amps indicates that the battery will eventually be discharged.

Since voltage is less costly to measure than current, it is obvious that voltage measurement is more common than current measurement.

To truly manage a power system, however, both voltage and current must be measured, because only together do you know absolutely what is happening in the electrical system. Without both voltage and current measurements, for instance, you can't determine when a battery is fully charged.

9.4 Where do we Measure?

Measurements taken in the wrong place aren't of much value. We are always amused when we see panel voltmeters wired up to measure the voltage at the panel, instead of wired to measure the voltage at the batteries. To the uninformed, it may seem like a trivial difference. After all, isn't the panel connected to the battery?

The difference is wiring and switch resistance. Every load turned on at the panel increases the current flow, and the wiring and switch resistance drop a greater voltage. What looks like 11 Volts at the panel may actually be 12 Volts across the battery, with 1 Volt lost to resistance between the battery and the panel.

We'd heard lame excuses for measuring panel voltage instead of battery voltage. The most common is the idea that it is important to measure the voltage available to power a device at its source of power, the panel. Bull feathers. If you want to know how much voltage is available at a device, then you should measure the voltage at the device, not some intermediate point between it and the battery.

The only proper place to measure voltage is directly across the batteries. Accept no other method!

Measurement of current is done in series with the battery. Current can be measured in series with either the positive or negative terminal. With present electronic devices, the best way to measure current is by using a shunt in series with the negative terminal, however. When placing a shunt in the negative lead of a battery, be sure that no devices are wired to the negative terminal, since they will bypass the shunt. Only current through the shunt will be displayed, so any bypass will be an unaccounted for current flow.

There are other ways to measure current than a shunt, such as magnetic "loops" containing Hall–effect sensors. While these devices offer isolation of the measurement circuits from the battery, and can therefore be placed in the positive battery lead easier, the loop devices are not very accurate, especially at low current levels. More information is presented below.

9.5 How We Measure

Sadly, many products are designed considering *consumer gullibility.* Electrical system instruments are not excluded. Marketing will argue that if the typical consumer can't tell the difference between an inexpensive instrument and one that costs more, then build the least expensive. We know better than to argue with marketing, but we think that in many cases, consumers can't tell the difference because they haven't been introduced to a difference. Just as power in language allows one to express complex ideas simply, exposure to highly functional instruments enlighten the consumer to the realm of what's possible. The enlightened consumer will always have a better chance to manage an electrical system.

Many factors contribute to the final functionality of an instrument, including:

- resolution ...how low a quantity can the instrument detect? Voltage should be measured to a resolution of 0.01 Volts. A resolution of current to 0.1 Amps is sufficient. It's easy to fake resolution without going to the expense of actually providing it. A display that shows 0.01 Volts may not provide that resolution. We know of at least one such instrument that has 0.05 Volts of resolution but is displayed as if it had 0.01 Volts. (Actual resolution is 1/5 of the display.)

- accuracy ...how close to reality is the displayed number? It's easy to provide digital displays, but not accuracy. Accuracy requires an expensive internal reference that is stable with both time and temperature.

- Temperature ...not only do internal references drift with temperature, but other components are also affected. Resistors and capacitors, two electronic components are both affected by temperature. A good designer makes sure that such components are not used in precision circuits unless temperature effects are cancelling.

- Time ... time takes it toll on electronic components just as it does everything else. Once again, a good designer attempts to use components so that their ratio is important, not their absolute value.

- Zero offset ... with an input of zero, what does the instrument display? A good meter will display zero, or provide a way of adjusting it so that it will display zero. Naturally, it must be stable at zero, or making the adjustment will be an ongoing exercise.

- sampling rate ... how fast are measurements taken? High sample rates are required to accurately track varying signals. On the average, current consumed by a water pump is constant, but current draw actually follows the load. That is, motor current will be highest at the compression point in the pump's stroke. Inverters draw current at 120 Hz, not pure DC. Wind generator outputs go up and down with wind gusts. Many instruments use too low a sampling rate to display accurate Amp–hour information. Worse yet, low sampling rates can *alias* with the frequency being measured and produce low errors at one time, and high errors at another, depending on where samples fall on the input signal.

How does a consumer determine resolution, accuracy, stability, and sampling rate? The best way to determine the first three capabilities is to compare the instrument to a known good digital meter, making sure to measure at the same point in the system as the point sensed by the monitor. Before you purchase an electrical system instrument, ask someone you know who has one if you can compare its results with your meter. Comparing the voltage readings is the easiest and a good starting place. If there is a large initial difference that can't be calibrated out, then beware. If you can calibrate the instrument to match your good meter, but it doesn't stay matched very long, then beware. If resolution is less than a good digital meter ... beware! If you can't trust voltage measurement, the easiest of all measurements, then you certainly can't trust any other function of the unit under test.

Determining an instruments sample rate will be hard to do without specialized test equipment. Here's where a written specification from the manufacturer is necessary. Anything less than 256 samples per second is too slow!

9.6 History of Digital Measurements

For over 100 years, analog meters with moving needles were used to measure voltage and current. The accuracy and resolution of an analog meter is poor, even under the best of circumstances. The quest for a replacement began in earnest in the early 1950s.

Digital voltmeters made their debut in the 1950s. Costing several thousand dollars, and weighing 50–75 pounds, the early digital meters weren't exactly drop–in replacements for analog meters. Yet digital meters offered unprecedented resolution and accuracy, and, as a result, unequalled ability to control industrial and technological processes. Control translates to dollars, so despite their high cost, there was a ready market for digital meters. Since then high quality digital meters have become a mass produced commodity. Only a good digital meter is suitable for DC measurement in an ample power system.

The early digital meters were constructed using telephone *stepping switches* which could be electrically pulsed to step from one position to the next. The switches were widely used at the time to make telephone connections automatically for early direct dialing systems.

Shortly after the stepping switch digital meters, came the reed relay meters. The logic driving the relays was tied to a comparator that sensed the input to be measured, and the logic balanced the input with a signal chosen by closing appropriate relays. By knowing which relays had to be closed to match the input voltage, the logic could also illuminate incandescent lamps indicating that input voltage.

In 1963, we joined a small team of engineers, draftsmen and technicians that was developing the first all *solid–state* digital meter. Though it operated logically like relay meters, transistors were used instead of the relays. Naturally, the solid–state meter was much faster, and without mechanical parts to wear out, offered higher reliability.

The first transistor meters used a balancing technique called successive approximation, where the input is successively balanced against a reference divider. The transistors selected what percentage of the reference voltage would be used as an approximation of the input. As the reading cycle progressed, the reference divider would come closer and closer to matching the input, until it would match. Typical resolution at the time was 1 mV, that is 1/1000 of a Volt, and several hundred readings per second could be made.

Other all–electronic means to digitally measure voltage were developed in the middle 1960s. First came the so–called integrating meter, closely followed by the *dual slope* integrating meter. While slow, the dual slope technique could be implemented very economically. It was also tolerant of electrical noise impressed on the signal being measured ... a considerable advantage. Most of the digital panel meters in use today use the dual slope integrating technique. Other, high speed analog to digital converters use one or more variations of successive approximation.

The 50 pound meters of the 1960's have been replaced by integrated circuits which cost a few dollars, making instrumentation a bargain, considering the high cost of quality batteries. No one operating a battery system should be without a high quality digital meter and the know–how to use it. If you wish to accurately measure inverter voltage you will need a meter that reads true RMS AC voltage. These meters do cost more than the typical meter that only measures average AC voltage.

9.7 DC Voltage Measurement

Voltage can be measured very economically with an analog meter. An analog meter operates around the principles of electro–magnetism. That is, as current passes through a wire, it generates a magnetic field that is proportional to the magnitude of current. The greater the current, the greater the magnetic force that is produced. Since current is proportional to voltage, voltage can be measured by the magnetic force generated. An analog meter operates around these basic principles. A small current is passed through a coil of wire, generating a magnetic force. The magnetic force deflects a needle that is calibrated to read in Volts. The needle is spring loaded so that with no force applied, the needle is at zero. As the magnetic force builds, the needle is affected, registering a level of voltage on a printed scale behind the needle.

How accurate can such an electro–mechanical device be? Not very. Friction, and parts length changes wrought by temperature changes make the analog meter inherently inaccurate. Operator interpretation of the needle position can lead to significant errors. The analog needle should always be viewed straight away, so its projection on the scale is directly under the needle. To view the needle from an angle is called parallax

Figure 9.1: Shunt, 50 mV at 200 Amps

error.

Initially, digital voltmeters cost more than analog meters, however, an analog meter can not register the true voltage of batteries with sufficient accuracy to manage them effectively. You will pay the price of digital instrumentation one way or another. The difference between a fully charged battery, and a dead one is only 0.6 Volts. Most analog meters have errors greater than that.

9.8 DC Current Measurement

Whereas the voltmeter has a series resistance that limits the amount of current through it, a current meter must offer very little resistance, so as not to interfere with the circuit under test. A voltmeter is attached across the circuit under test, but the conventional ammeter is attached in series with the circuit to be tested. To make a test measurement of current, the circuit must be temporarily broken and the meter inserted in the break ... more than a minor inconvenience.

For permanent measurement of current, a shunt is wired in series with the conductor carrying the current to be measured. A shunt is a precision low resistance device. Since the resistance of the shunt is known, current can be determined by measuring the voltage drop across it, and applying Ohm's law. Shunts are constructed as one or more manganin strips soldered into brass end pieces. Manganin is an alloy of copper, manganese, and nickel which exhibits very stable resistance over temperature extremes.

Shown in Figure 9.1 is a shunt that develops 50 millivolts at 200 Amps.

All shunts are not created equal, however. The higher the resistance of the shunt, the more voltage it drops, and the easier it is to measure that voltage drop. At first glance, this would appear to be beneficial. Unfortunately, the higher the voltage drop in the shunt, the more the power dissipation, and the less voltage left to power your equipment. Also, high power dissipation in the shunt leads to premature failure. Faced with these dilemmas, some manufacturers opt for ignorance ...yours. They count on you not knowing that high voltage shunts waste power and can in fact be a fire hazard. High voltage shunts have another problem ...they often burn out when engine starter currents are conducted. A high voltage shunt is one which develops more than 50 millivolts at its rated current, i.e. 100 millivolts at 100 Amps.

If you are trying to measure 200 Amps, a proper shunt would develop 50 millivolts, and no more. That's fine at high currents, but a proper low voltage shunt has less sensitivity to low currents. It is important to have a resolution of 0.1 Amp. Getting this resolution with a high voltage shunt is much simpler and less expensive circuits can be used.

If a proper low voltage shunt is used, then obtaining a resolution of 0.1 Amps requires a sensitive electronic circuit ...costly, of course. What all this boils down to is complexity that the concerned manufacturer faces up to by designing the necessary sensitive circuitry.

A recent ammeter development is the *clip–on* ammeter. The clip–on has an opening jaw that is placed around a wire carrying current. The flowing current causes a magentic field, the force of which is proportional to the amount of current flow. The clip–on meter actually measures the strength of the magnetic field with a *Hall effect* sensor.

The Hall effect was discovered in 1879 by E.H. Hall[1]. It remained a laboratory curiosity until semiconductor technology could develop materials that generate a Hall voltage several times that of ordinary metals. In a current carrying conductor, a magnetic field is generated perpendicular to the wire. Hall discoverd that when an external magnetic field was applied to a current carrying conductor, the wire's magnetic field would be skewed. This in turn causes the electric charges in the wire to separate such that a voltage is generated from one side of the wire to the other side. The amount of voltage that is produced is dependent on the amount

[1]Edwin Herbert Hall, American physicist, 1855–1938

of current flow.

While Hall effect sensors are useful as clip–on ammeters for testing purposes, the standard resistive shunt is best suited for permanent installation. The Hall sensor tends to exhibit drift over time and temperature, and absolute accuracy is not as great as can be achieved by the shunt and precision electronic circuits. Hall effect sensors also require a significant amount of power to operate, making them impractical in a conservative energy system.

9.9 Power Measurement

As we learned in prior sections, power is measured in Watts, and is the product of Volts and Amps. If you are presently measuring both voltage and current, then you can easily determine power consumption with arithmetic. On the other hand, power consumption can be measured directly by using an electronic multiplier that produces a voltage output equal to the product of voltage and current. Some modern instruments, such as the Energy Monitor/Controller from Ample Technology[2] also calculate kilo–Watt–hours, kWh, for the user.

9.10 AC Volts Measurement

AC voltages are usually measured by simply rectifying the AC to a DC voltage, and then the DC voltage is measured. Rectification in a low level circuit produces the average value of the AC.

Scaling is provided in AC meters so that the average value of the AC is displayed as an RMS value. You may recall that the average DC produced from rectifying AC is .637 of the peak AC voltage. For the same peak voltage, the RMS value is .707. The RMS value is an accurate measurement of the equivalent DC power in an AC voltage, so it is preferred over the average value. Average reading meters are scaled by a factor of 1.11 to display RMS.

Scaling the average value to display RMS works on a pure sine wave. For shore power, or generators, the sine wave is generally pure enough that significant errors are not incurred. Average reading meters fall short

[2] Ample Technology, Seattle, WA.

when connected to inverters. Inverters, as a rule, don't produce a sine wave, but rather a rectangular pulse. The width of the pulse varies to maintain a constant RMS voltage. The average reading meter can not accurately measure a rectangular pulse ... as the pulse width changes, so does the reading, even though the RMS value stays constant. Errors as large as 30% are typical.

Measurement of true RMS voltage is more expensive to perform than an average reading. Fluke[3] now produces a handheld Digital Multimeter, Model 87, that measures true RMS. It also measures frequency and pulse width, and other functions, so a direct comparison in price with their average reading multimeters is not possible, but the Model 87 costs about 3 times the Model 77, which does not measure true RMS.

AC voltage measurement is not generally very critical. An analog meter is sufficient for sine waves, but is not accurate on inverters. Low or high voltage takes its toll on motors and other electronic devices, so some sort of instrument is necessary. AC voltage should be between 105 and 130 Volts.

9.11 AC Current Measurement

AC current measurement is most often done with an analog meter that is connected directly in series with one leg of the AC wiring. Such a measurement is simple and effective, although it does place the meter at AC potential.

For isolated measurement, a current transformer can be used. The analog meter may still be used.

A shunt can also be used to measure AC current, but since it costs more than the current transformer, and offers no isolation, a shunt is only used when precision is necessary. The voltage developed across the shunt must be amplified, and then it can be measured with either average or RMS methods.

9.12 Amp Hour Measurements

We're a society accustomed to fuel gauges ... how much remains? We

[3] John Fluke Manufacturing Company, Everett, Washington.

don't want to 'guess what', and will often trust a poorly designed instrument over our intuition. Measurement of Amp hours consumed and especially Amp hours remaining is fraught with difficulties. If we want to 'know what', then we must arm ourselves with a little knowledge about the process of measuring Amp hours.

Before we can measure Amp hours, we have to measure Amps. Naturally, if we are going to measure Amps, it would be a waste not to also display the actual Amps read. As mentioned, current flow in Amps is best measured by placing a *shunt* in series with the flow and measuring the voltage developed from one end of the shunt to the other. The voltage measured is directly proportional to the current flow.

9.12.1 What is an Amp Hour

Let's ask a simple question. What is an Amp hour? It isn't a *fact* that an Amp hour is 1 Amp for 1 hour. That happens to be a definition. A fact is something that holds true based on a definition. Likewise, 1 Amp for 100 hours is defined to be 100 Amp hours, the same as 100 Amps for 1 hour. It all seems simple enough, just multiply Amps times hours to get Amp hours.

Unfortunately, an Amp hour is not an Amp hour in a practical sense. A battery that will supply 1 Amp for 100 hours will not yield 100 Amps for 1 hour. If the battery is an old fashioned thick plate deep cycle unit, it will only yield about 45 Amp hours at a discharge of 100 Amps. Stated in other terms, a 100 Amp hour battery is only a 45 Amp hour battery at 100 Amps. So what is an Amp hour? It seems like such a simple question. An Amp hour is defined to be the product of Amps and time, but in practice, Amp hours are different if they are consumed at different rates. Of all the facts that bear on the subject of Amp hour measurement, this fact must be well understood before a proper technique for Ah measurement can be realized.

Let's ask the simple question again. What is an Amp hour? A battery that will yield 100 Ah at 77° F will only yield about 80 Ah at 32° F. On the other hand, it will yield about 110 Ah at 122° F. To be meaningful then, measurement of Ah must compensate for temperature.

So what is an Amp hour? Consider this. One morning your car won't start. You crank the engine until it will no longer crank. After an hour rest you decide to try one more time. Lo and behold the engine cranks just great and the engine even starts! The magic at work is none other than the so called *recovery* phenomena. A battery which has been discharged at a fast rate can be rested and regain capacity. Whatever an Amp hour is, it is not something *basic*. We can not simply measure Amps and multiply times time to derive meaningful data about the Amp hours remaining in the battery. Such a measurement is *linear*, and battery discharge is *exponential*.

9.12.2 A Linear Amp Hour Meter

To make a linear Amp hour meter, requires a current shunt to develop a voltage. As mentioned, a manufacturer of high quality equipment would use a low voltage shunt. Depending on what kind of shunt used, there may or may not be signal conditioning circuits for the shunt.

The voltage across the shunt is proportional to current. By converting the voltage to a frequency with an inexpensive integrated circuit, the current is now proportional to a frequency. By counting the frequency with a precise time base, we have effectively counted the Amp hours of consumption. With a minimal amount of scaling, the Amp hours consumed can be displayed digitally. Scaling the frequency is not without its difficulties ... all sorts of circuit components drift with time and temperature to force frequent re–calibrations, typically before each use.

Amp hour meters based on this principle have been widely used in the past for battery testing. By holding both the current and temperature *constant*, a battery manufacturer can measure, and then rate their batteries in reserve minutes, i.e. 25 Amps for 300 minutes.

A linear Amp hour meter is fine as long as both current level and temperature remain constant but life afloat is rarely this simple.

If we can count linear Amp hours leaving a battery, why can't we count them going back in and know when the battery is fully charged? All we need to do is add a little fudge factor for recharge efficiency and sure enough, the meter will read zero when the battery is fully charged ... sort of.

In general, recharge efficiency is in the range of 5–25%. That is, you must put 5–25% more Ah into a battery than what you withdrew. There's a lot of room for error in 5–25% ... that's the good news. The problem is worse than it appears. Recharge efficiency is highly dependent on the rate of the prior discharge. It is also dependent on the discharges that have occured previously, the rate at which the battery has been recharged, and the rate at which it is presently being recharged. The issue of recharge efficiency is so complex that even a supercomputer would be hard pressed to model all the interactions that occur. With recharge efficiency so difficult, it is extremely unlikely that a potentiometer adjustment could ever be set correctly unless a very typical discharge and charge regimen is followed repeatedly. The economy of such a meter may make it useful as an indication device, however.

9.12.3 An Exponential Amp Hour Meter

Measuring Amp hours in a realistic setting means that we must account for temperature and the rate of discharge. Correcting for temperature is fairly straightforward, but accounting for the rate of discharges is non-trivial. Perhaps this explains the dearth of true Ah meters on the market.

Back in 1897, a battery researcher by the name of Peukert published an equation that allows the calculation of available capacity at any discharge current. His equation, still valid today, states:

$$I^n t = C$$

In Peukert's equation, C is a constant, namely the capacity of the battery. The letter I denotes the discharge current in Amps. The exponent n is related to battery construction, and is also a constant for any given battery. The letter t represents the time of discharge. Exponent n varies from more than one to about 2. Exponent n can be calculated by knowing the capacity obtained at any two discharge currents. That is, if we first discharge a battery at 5 Amps, and later at 20 Amps, and calculate the Ah obtained for each discharge, then the exponent n can be calculated from the equation below.

$$n = \frac{log t_2 - log t_1}{log I_1 - log I_2}$$

In the equation above, t_1 is the hours of discharge at the current I_1. The second discharge is t_2 and I_2.

It seems simple enough. Discharge a battery twice, calculate exponent n, and thereafter use a computer to evaluate Peukert's equation in *real time* (at least once a second) to determine the instantaneous capacity consumed. By summing the consumption of each second, total Ah consumed becomes a known. There's at least one fly in the ointment. Evaluating Peukert's equation requires a fair amount of computer power ... beyond the capability of most low level micro–computers, and a complex programming exercise on any computer.

There is an even more subtle problem related to the recovery phenomena. If you discharge a battery at a high rate, and then let it rest, it is possible to continue discharging at a low rate. In the end, you may get almost as many Amp–hours as you would have, had you discharged it at the low rate continuously. Plate thickness, as usual, dictates the final outcome. While Peukert's equation offers much insight into the problem of computing Ah consumed, can it be applied? The recovery phenomena prevents the direct application of Peukert's equation when the load current is highly variable. Such is the case when load current consists of some combination of autopilot, running lights, bilge pumps, inverter/microwave discharges, and many other variable and time dependent loads.

To use Peukert's equation requires that we know or calculate his exponent n. While this is possible on each battery to be monitored, it can also be accomplished on a class of batteries. For instance, manufacturer Y will probably always have about the same size exponent. Classifying manufacturers by their exponent is quite feasible and presents the fewest problems for the user.

Whereas the linear Ah meter suffers from time and temperature effects, the clock frequency used by the computer in an exponential meter, is fixed by a crystal that maintains great accuracy under variations of time and temperature. A crystal is said to *age*, as its frequency shifts slightly with time. Aging is somewhat predictable, so built–in correction of internal time is possible, but probably not necessary.

9.12.4 An Energy Monitor/Controller

Since computing power is essential to make accurate Ah measurements, it makes sense to spread the cost of the computer over a wider functionality. The Energy Monitor/Controller from Ample Technology[4] not only provides voltage, current and Amp–hours consumed/remaining, it also per-

[4] Ample Technology, Seattle, WA.

Figure 9.2: Energy Monitor/Controller

forms solar/wind control and load disconnect functions. To compute Ah remaining, the Energy Monitor/Controller use Peukert's equation that accounts for the rate of discharge.

One of the most important functions of the Energy Monitor/Controller are the multiple alarms which can be individually enabled or disabled. Alarm setpoints are programmable. Sudden failures of electrical systems are not common. It is much more common to suddenly discover a problem with the system because no alarm preceded the ultimate failure. An electrical system alarm can save your life!

Figure 9.2 shows an Energy Monitor/Controller, (formerly Energy Monitor II), made by Ample Technology that monitors two batteries, and an alternator or solar/wind generators. It employs Peukert's equation to determine Ah remaining.

9.13 Small System Instrumentation

For a small system with a single house bank of moderate capacity, instruments for voltage and current are sufficient with the exception of determining a depth of discharge of 50%. With some sort of record keeping regarding average Amps of load, it may be possible to calculate an approximate depth of discharge. As long as a reasonable margin of error is allowed, satisfactory battery performance can be expected.

With a slightly larger system that employs two house banks, depth of discharge can be determined by accurately measuring the rested voltage. As long as several days of capacity are available in the two banks, battery management will be simple and effective.

9.14 Large System Instrumentation

Like the small system, measurement of both voltage and current is necessary. In a larger system, Amp hour measurement is also required.

Those who take the time to add up their Ah consumption usually find that passage making under sail consumes much more than swinging on the hook at anchorage. Power yachts have plenty of power during passages, but generally have high demands at anchorage because of refrigeration and ice making equipment. In either case, the extremes of demand can probably not be met without daily recharges because battery capacity for several days at the maximum rate of consumption would mandate monstrous battery banks.

When frequent recharging is required, a rested battery is not an option. The only way to keep track of depth of discharge in this situation is with an exponential Amp hour meter. The linear Amp hour meter is guaranteed to be in error, when current demand is highest. When you need Ah computation the most, linear measurement exhibits it greatest errors.

9.15 Summary

You can't see electricity ... to manage an electrical system, you must have accurate instrumentation. Volts measurements are handy, but current measurements provide information that voltage alone can't.

Analog measurements are little better than idiot lights. Digital readings are required. Measurements should be taken at the batteries, not the electrical panel.

Current measurements are best done with a low voltage shunt that develops only 50 milli–Volts at 200 Amps.

Linear Ah measurements are most easily accomplished, but do not account for rate of discharge. To provide true Ah measurement requires that Peukert's exponential equation be evaluated in real time.

Chapter 10

System Design

10.1 Introduction

There's nothing quite as satisfying as owning an electrical system that performs as intended. While others nearby always seem to have problems with batteries or refrigeration systems, the owner of a performance system enjoys the fruits of a good design. Food and beverages stay cold, electronic devices operate as expected, and adequate lighting illuminates the night.

Performance electrical systems don't happen by accident. The first rule is conservation ... without conservation, no electrical system can be expected to perform well. The second rule is to waste as little energy as possible charging batteries. Running engines to charge batteries is particularly expensive when lots of horsepower is driving a puny alternator, or a large generator is driving a low output battery charger.

To design an electrical system to fit your needs requires that you understand how to fit the pieces together which have been presented earlier in the book. The key concept is balance. Battery capacity must be balanced with load demands; charge capability must be balanced with battery capacity.

10.2 Decide on 12 or 24 Volts

A 12–volt house system is generally best for most boats and RVs which will probably need some 12 Volt power for running loads. Mixing systems with both 12 and 24 Volts can be done successfully, but it entails extra equipment and complicates management of the system.

While it's true that 24–Volts in more efficient because there is less power lost in the system wiring, initial expense will be higher since 24–Volt devices cost more. We think that it is more prudent to stick with a 12–Volt system and make an initial investment in larger gauge wires for the main power handling connections.

We've seen both boats and RVs with mixed systems because the owner got some bargain on a piece of 24–Volt equipment. That bargain turned out to be part of an ongoing problem with management of the system ...running both a 12 and 24–volt alternator in one instance.

To some extent the arguments for 12–volt systems extends to remote homes, but more 24–volt systems are used there. With the new sinewave inverter/charger from Trace[1], which is only produced in 24–volts, such a system makes sense. Rather than wire the house for DC appliances, it can be wired as normal for AC, which is provided by the inverter. The Trace inverters can be stacked to provide 220 VAC, useful for some home appliances.

10.3 Determine the Loads

System design proceeds by determining what the loads on the system are, and how much energy they require. Once loads are known, other aspects of the system can be determined.

Determining the loads means to calculate how many Amp–hours are consumed by the loads on a daily basis. Different loads are used at different times. Its not likely that an autopilot is used at anchorage, for instance. Because of this, you will probably need to consider several scenarios, calculating the daily Amp–hour load for each of them.

[1]Trace Engineering, Arlington, WA

10.3.1 Refrigeration

Since refrigeration may be the largest load on the system, pay particular attention to how you will refrigerate and how much space will be cooled. As mentioned, a DC unit with holding plates is the most efficient in the long run. Holding plates substitute for battery capacity whenever the plates can be pumped down at the same time charge sources are operating.

DC refrigeration operating via a thermostat from the battery is also efficient for small refrigerators. We replaced a propane refrigerator in our camper with a same size DC unit and have been more than pleased with it's performance even during 100°F weather. Only in the silence of a pristine campspot can the unit be heard to purr, and with limited propane reserved for cooking, concerns about running out during an outing have dissolved.

While we prefer DC refrigeration in boats and RVs, propane refrigeration may be a good choice for remote homes where alternate energy sources are limited. If you aren't going to be able to get enough energy from the sun or wind to power refrigeration then by all means consider propane. Be aware that propane operated refrigerators are very inefficient, and be prepared for the extra fuel expense.

What ever method of refrigeration is used, the daily Amp–hour consumption must be calculated.

10.3.2 AC Power from Inverters

In many systems, AC power is desired to operate microwave ovens, toasters, hair dryers, coffee makers, and tools. Extensive use of these devices would overpower many systems, however, used judiciously, such conveniences can be accommodated.

Total Amp–hours used by an inverter may not be much, but high currents can result. A 1500 Watt load on an inverter may be a 1700 Watt load on the batteries by the time inverter efficiency is accounted for. The result is a discharge of about 150 Amps. According to a rule of thumb presented earlier, about 600 Ah of battery capacity is necessary for this load. As mentioned, some gel batteries supply high current readily, and so capacity can be reduced.

10.3.3 Lighting

Lighting can represent a fair load, especially if incandescent lights are used. It's a normal tendency to underestimate total hours that lights are on, forgetting about the times that lights are unintentionally left on.

10.3.4 Fans for Cooling and Heating

Fans used for ventilation may be run many hours in hot climates. Likewise, fans in diesel or other heaters may run continuously at times. A 3 Amp fan running for 24 hours is a sizeable load.

10.4 Determine Battery Capacity

Once a tabulation of all the loads is done, battery capacity can be determined. Amp–hour capacity should be four times the largest draw in Amps, or four times daily Ah consumption.

You don't have room for all the batteries needed? Besides reviewing consumption to see where conservation can be practiced, there are other choices that can be made. First of those is to consider sophisticated instrumentation. Contrary to many people's belief, the less the available capacity relative to demands, the more insturmentation is required. If you're going to cut things close, you need to know what close is!

As mentioned, some gel batteries are not as sensitive to regular discharges below 50%. If space allowance limits battery capacity, then use of gel batteries should be considered. Some gel batteries also provide more Watt–hours for the same number of Amp–hours, which allows less Amp–hours to be available.

Conservation can be practiced easier if you have an instrument that provides a high current alarm, such as the Energy Monitor/Controller from Ample Technology[2]. With such an instrument a current budget can be programmed, and an alarm sounded whenever the budget is exceeded. (The alarm can even be used when the stereo is turned up too high.)

Juggling space and battery volume is usually an exasperating exercise. Batteries are never made in exactly the right size for the available space. In general, you want to select the biggest battery size available. This will

[2]Ample Technology, Seattle, WA

result in least cost, and a more robust battery bank. For instance, an 8D battery is more ruggedly constructed than a Group 27 size ... even though three of the Group 27 units will have slighty more capacity than an 8D, they probably won't offer as many discharge cycles.

10.5 Select a Battery Configuration

With battery capacity determined, how will the capacity be distributed? Prior to the introduction of inexpensive meters that display Amp–hours remaining, it was natural to divide battery capacity into two equal size banks. This permitted resting a bank for 24 hours in order to use voltage as a determinant of capacity remaining.

Concentrating all the capacity on one bank, however, has much merit. This configuration is preferred when an Amp–hour instrument is used. Merits of the single house bank are:

- A single house bank suffers less from Peukert's effect than two banks that are alternately discharged.
- Watt–hours provided by a single house bank are greater for any given Amp–hours over the Watt–hours provided by two banks of the same total capacity.
- A single house bank is easier to wire and instrument.
- A single house bank is much easier to operate than two banks where decisions about which bank to use must be made.
- A single house bank can be used without a battery selector switch ... a common source of problems and mis–use.

10.5.1 Charging the Second Battery

Most marine and RV systems will use a dedicated starter battery as insurance for the day that the house bank is dead. The starter battery will need to be charged. In a two bank house system, some means of charging both banks is required.

How should the second battery be charged?

The least desirably way is to use an isolator, or charge splitter as they are often called. Isolators are diodes that conduct charge to more than one battery, while keeping the batteries from discharging into each other as they are loaded. Diodes drop a voltage, so unless you have a remote sensing regulator, batteries will not receive a full charge. Most remote sensing regulators have a single sense wire. Which battery does it go to? The one that it doesn't connect to will eventually overcharge as it charges to the peak of any ripple on the alternator or charger output.

A manual switch can be used to short both batteries together during a charge. This is a nuisance, and if it's forgotten, then the starter battery will be discharged along with the house bank.

An automatic solenoid that connects the two battery banks together is an effective solution if properly implemented. Voltage sensing alone may result in solenoid chatter at times. Solenoids draw a lot of current, usually making them ineffective for solar and wind applications. Solenoids that magnetically latch into the closed state are available, however. Latching solenoids don't require current to maintain a connection, so they can be used with solar and wind chargers.

Another way to charge the starter battery is the Eliminator, or battery two regulator made by Ample Technology[3]. This patented device connects between a house and starter battery and siphons charge current from the house circuit whenever it is being charged. The Eliminator regulates the voltage applied to the starter battery and even includes temperature sensing to provide a correct charge as daily or seasonal temperature changes take place. The Eliminator consumes very little power, allowing its use with any charge source attached to the system, including small solar panels.

10.6 Select Ample Instrumentation

There's no hope of getting good battery performance and life without instrumentation. As a minimum, house battery voltage, current and Amp–hours consumed is necessary. The BatMon manufactured by Ample Technology is able to display those parameters.

A more completely instrumented system would include house battery

[3] Ample Technology, Seattle, WA

temperature, starter battery voltage, Amp–hours remaining, solar and wind current as well as Amp–hours produced, total Amp–hours consumed over the life of the batteries.

As mentioned, alarms are a very important part of the instrumentation system ... the more the better!

10.7 Select Charge Regulation/Control

Battery charging must be tightly controlled and temperature compensated for battery longevity. This means that alternators should be regulated with a multi–step regulator with an external temperature probe that mounts onto the battery. Battery chargers should likewise be equipped.

Solar panels and wind generators can be regulated, or switched on and off as a strategy for maintaining the batteries at a high state–of–charge. Typically, a remote home won't have enough charge capacity to maintain a full charge long enough to warrant a regulator. However, on/off or *bang–bang* control does allow some unnecessary cycling of the batteries. If you are lucky enough to have plenty of solar and wind charge, then a regulator as opposed to a controller should be considered.

If a regulator is used, it should also provide multi–step performance with temperature compensation. A controller should also have its disconnect voltage temperature compensated. That is, the voltage at which the controller disconnects the solar and/or wind charger from the batteries should be compensated for battery temperature, opening at a lower voltage when temperature is higher.

10.8 Summary

In this chapter the subject of 12 and 24 Volts was presented. How to determine the system load was discusssed with the load's impact on necessary battery capacity.

Instrumentation and regulation/control of the charging sources was presented with recommendations for proper levels of instrumentation and control.

Chapter 11

Electrical System Components

11.1 Introduction

In this chapter, we consider the many components of a system that aren't batteries, loads, sources, or wire related. Components such as fuses, fuse holders, switches, breakers, relays, transformers, electrical panels, connectors, isolators, snubbers, diodes, and regulators are covered.

11.2 Fuses

There are many shapes and physical sizes of fuses, and fuses in every imaginable rating from 1/500 Amp to 100,000 Amps. All fuses are basically a metal link that melts when too much current flows through the fuse. Some fuses, called dual element fuses, have a spring and a metal link. Fuses are made for different applications. Some fuses are designed to blow only after a long sustained overload. Others blow quickly when overloaded. There are intermediate grades, from very fast acting to long time delays. If you go into an electronic store and ask for a 10 Amp fuse, you may get asked whether you want a fast blow, or a slow blow fuse.

For the most part, you will use time delay, slow blow fuses. This type of fuse is for basic wiring protection ... under a short circuit, and before the wire gets hot, the slow blow fuse will interrupt the current flow.

Fuse Type	Size (inches)
5 x 20mm	3/16 x 3/4
1AG	1/4 x 5/8
3AG	1/4 x 1 1/4
4AG	9/32 x 1 1/4
5AG	13/32 x 1 1/2
7AG	1/4 x 7/8
8AG	1/4 x 1

Table 11.1: Electronic Fuse Data

Fast blow fuses are used whenever the circuit can not sustain an overload very long. Very fast acting fuses are used to protect power semiconductors. Fast blow fuses will not work in many circuits, such as pump motors, or electronic gear. When power is first applied to a motor or power supply, there is an initial inrush of current. A fast blow fuse, rated for the steady state draw of a motor will blow on inrush current. On the other hand, slow blow fuses don't go fast enough to protect semiconductors, so don't substitute a slow blow where a fast blow is specified. All fast blow fuses are not alike in operating speed. Many are not fast enough to protect semiconductors, so when a manufacturer has specified a particular very fast acting fuse, heed the specification.

Fuses are rated according to nominal load current, and will open when current exceeds 135% of the rating. This parameter is temperature dependent. At 125° F, a typical fuse can only handle about 75% of its rating. Corrosion on a fuseholder usually results in heat generation, and the rating of the fuse is accordingly less. If a given size fuse that has been working starts failing often, check for heat buildup at the fuse.

Fuses do have voltage ratings. When a fuse opens, it must be able to sustain the full circuit voltage without arcing. Typical voltage ratings include 32, 60, 125, 250, 300, 500, 600 and 750 Volts. As long as the current interrupting characteristics are the same, it is permissible to use a higher voltage fuse in low voltage circuit, but not vice versa.

Given in Table 11.1 are types and sizes of electronic fuses covering the range of 1/500 to 30 Amps. The most popular sizes are the 5x20 mm,

Figure 11.1: Fuse Symbols

3AG, and 5AG. The AG is historical baggage that stands for *automotive glass.* Fuse type and rating is stamped on the fuse.

Naturally it is a good idea to have spare fuses. Equipment manufacturers are usually willing to supply fuses for their products, but sometimes the cost of handling the fuse order makes for an expensive fuse. It is better to make an inventory of all the fuses used, and visit an electronics supply store. You may have to purchase 5 fuses in a box, but the spares can always be traded later. Or, you can get together with a few friends and split the fuses as needed.

Fuses appear to be sealed, but they are only designed to keep dust or lint from entering. If they fill with water, you must replace them.

Figure 11.1 shows the schematic symbol for a fuse.

11.3 Fuse Holders

There are several types of fuse holders that are applicable to marine application.

Panel mounted holders have a threaded body with a nut that allows the holder to be secured to a panel. The fuse is replaced from the front by unscrewing a cap. Some caps loosen with a 1/4 turn locking action. Panel mounting fuses are available in waterproof versions.

Fuses can also be mounted in fuse blocks. Fuse blocks do not enclose the circuits, so they should only be used inside enclosures that provide protection. Fuse blocks are more economical than panel holders, particularly the blocks which hold multiple fuses. Connections can be made with quick connect tabs, soldering, or with ring terminals and screws.

A third style of fuse holder is the in–line variety. They are available with pigtail wires which require a wire splice to connect, or with crimp ends which accept a wire directly. Waterproof in–line fuse holders are available.

11.4 Breakers and Fuse Switches

There are several types of circuit breakers. The least expensive types are bimetal contacts that open up if they get warm enough. They act just like a thermostat, but are designed as current interrupters. Other circuit breakers operate on magnetic principles, opening when a specified current flows. The flowing current generates a magnetic field, and when the field strength reaches a predetermined level, a holding latch is tripped, opening the breaker. As you might expect, the magnetic breaker is more expensive, but also more repeatable and faster acting.

Circuit breakers are made for AC or DC circuits. Combination AC/DC breakers are also available, and widely used in the marine industry. Like fuses, circuit breakers are specified with a nominal current and will open at 115% of nominal. And, like fuses, circuit breakers can be slow or fast operating. It should be noted that the combination AC/DC breakers are slow operating, which is alright for most applications since the prime purpose of a breaker is to protect wiring from burning up under short circuits.

Because of inrush currents, a breaker sized too close to the nominal current draw of a motor may open when the motor is first actuated. A larger size breaker will solve this type of problem, but it may fail to give adequate protection. Selecting a breaker with a longer time delay is the better solution.

There are three types of handles on breakers; push to reset, toggle handle and rocker handle. The push to reset type of breaker is not used as an on/off switch, but is strictly a breaker. Once the breaker is on, it remains on until an overload trips it. The toggle handle breaker is sometimes called a bat handle breaker. Whatever it is called, it is easy to knock off when trying to turn another breaker on or off. We hit a reef when the autopilot breaker got knocked off by accident ...it might have been a disaster, except we went over the reef into deeper water. Naturally, the marine industry has pretty much standardized on the toggle handle, when the rocker handle makes far more sense. The rocker handle is not nearly as prone to being accidently knocked off when reaching for another circuit. When it is your own boat, and you accidently trip a breaker, it is easy to turn the switch on again. With friends aboard, the chances are good that one of them will trip a breaker and not even know that they

Figure 11.2: Breaker Symbols

did it.

It has become pretty much standard to have a circuit breaker on nearly every conceivable circuit. This practice is not only expensive, but it also consumes precious panel surface. We repeat ourselves, in saying that a breaker is really only there to protect wiring. Often, breakers can be wired to protect several circuits, with switches utilized to turn individual circuits on and off. A newcomer to the market is a fuse/switch which has a fuse internal to the switch. The fuse is replaceable from the front of the switch, by turning the switch handle 1/4 turn. These switches can be mounted in much less panel space than the circuit breaker, so smaller panels can be used to serve the same number of switches. A few circuit breakers can still be used to feed power busses. For instance, you might have a power buss that feeds the port side, and another that feeds the starboard side. Yet another breaker might control bilge circuits, and a fourth breaker control topside functions. Individual circuits can be turned on or off using regular toggle switches, or, for ultimate protection, the fuse/switch.

Figure 11.2 shows some of the symbols used to represent breakers on schematics.

11.5 Switches

11.5.1 Battery Selector Switches

Somewhere back in the dark history of electricity aboard boats, someone decided that two battery banks were sufficient. Indeed, with proper instrumentation, including Amp hours consumed and remaining, two banks are sufficient. A two battery selector switch which can select either battery, or both is also easy to manufacture. The old 1, 2, both switch became the standard selector.

But, batteries were not instrumented sufficiently, and so they were

Figure 11.3: Single Pole, Single Throw Switch

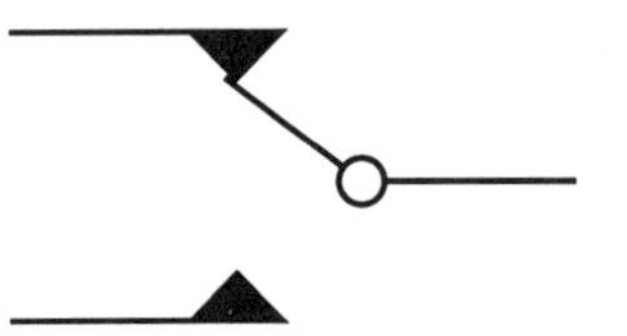

Figure 11.4: Single Pole, Double Throw Switch

often flat just when the users needed power the most. This meant that some people started carrying an extra battery bank. With three batteries, the 1, 2, both switch wasn't exactly applicable, but with two of them, one could select any or all of three batteries. In the chapter of schematics, a diagram shows how three batteries can be wired with two selector switches of the 1, 2, both variety.

The obvious solution to a three battery system is one switch per bank. Turn a battery switch on, and that battery bank is put into service. With all switches on, all batteries are connected in parallel. Switches for each battery increases the risk that all switches may be open. This could be a disaster if the alternator were running. Diode isolators should be used to charge multiple batteries where the risk of opening all batteries is high.

11.5.2 Toggle Switches

Toggle switches are the simple on/off switch with a long stem handle. A rocker switch may serve the same function. Some toggle switches are made that come with two contacts. The contacts are connected in one position of the handle, and open in the other. Such a switch is called a single pole, single throw switch, abbreviated SPST. A schematic symbol is show in Figure 11.3.

More often, you find that a toggle switch has a center contact, and two contacts that make to the center contact depending on the position of the toggle handle. The center will connect to one contact with the handle in one position, and to the other contact when the handle is in the other

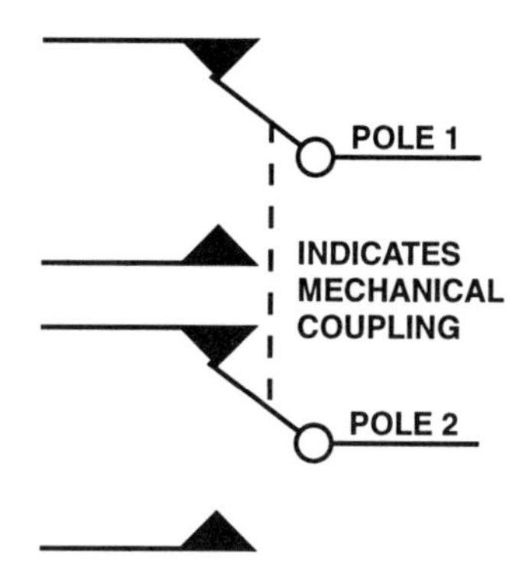

Figure 11.5: Double Pole, Double Throw Switch

position. This type of switch is called a single pole, double throw switch, abbreviated SPDT. Refer to Figure 11.4 for its schematic.

A toggle switch may have more than one pole. That is, it has a second or third (or fourth, etc.) center contact that may make to either side of the switch. Each circuit is independent, so the switch can be used simultaneously with different voltages. A double pole, double throw, (DPDT) switch is shown schematically in Figure 11.5.

It is a good idea to carry a spare switch of different varieties. Sometimes you may wish to add something, and need a switch to control it. When we spent a summer in Mexico, we added several fans to the boat. The only switches we could find locally were made for household use. Fortunately, we had a few spare switches that were pressed into service for the fans.

By all means, carry a heavy duty toggle switch. It may find use as a switch for an extra bilge pump, or it may be used to control charging from an alternator. Refer to page 86 of *Living on 12 Volts with Ample Power* for more details about using a switch as a temporary replacement for a regulator.

11.5.3 Rotary Switches

Rotary switches come in a variety of configurations, from a single pole switch on a single wafer or deck, to multipoles on a single deck, to multipoles on multiple decks. A wafer or deck is a flat circular plate that

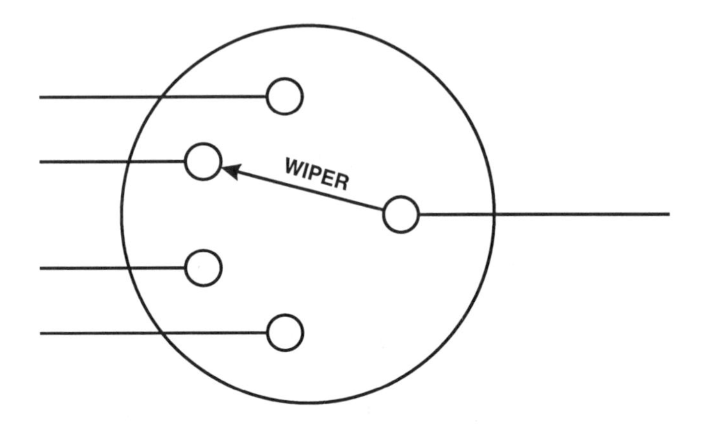

Figure 11.6: Rotary Switch

has contacts arranged around its periphery. The wiper, or contact that makes to other contacts as the switch is rotated, is in the center of the wafer. Each wiper is a pole, or common element. There may be several poles on a single wafer, depending on how many different functions need to be switched. By adding wafers, or decks, as many poles as desired can be added to a rotary switch. A schematic symbol for a rotary switch is shown in Figure 11.6.

Rotary switches are normally detented such that the switch comes to rest at a contact. Detents are made with a spring loaded ball that snaps into position when the switch is rotated close to a detent position. Sometimes detent mechanisms break and it is then possible to leave the switch between positions.

Rotary switches come in shorting or non–shorting varieties. Shorting switches short between adjacent positions as the switch is rotated. The typical 1, 2, both battery selector switch is a shorting type rotary switch. By shorting between positions, a circuit is never open. In the case of the battery selector switch, an open circuit between positions can result in a damaged alternator if the alternator is running when the switch is rotated.

Shorting switches are not desirable in all circumstances. For instance, a switch that selects a measurement channel for battery voltage should not short between the batteries as it rotates. Such a light duty switch could easily be welded in position if heavy battery currents were to flow

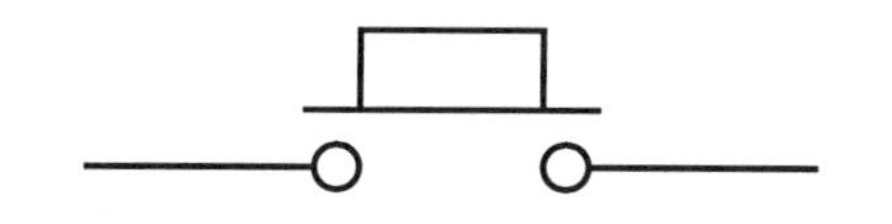

Figure 11.7: Push Button Momentary Switch

when the switch shorted.

Not all rotary switches have a contact for every detent position. Some are binary coded. For instance the selector switch in the Ample Power Electrical System and Amp Hour Monitor has 16 positions, but only five contacts on the switch. The center contact is the wiper, and it makes to the other contacts in a binary code as the switch is rotated through the 16 positions.

11.5.4 Momentary Switches

Momentary switches are those switches that make contact only as long as you hold the switch. Momentary switches are best known as the push button variety, although momentary switches come as toggle or bat handle units, and even as rotary switches. The key operated starter switch is a good example of a spring loaded rotary switch.

If a circuit is equipped with a momentary switch, it is not good practice to activate the momentary circuit too long. The switch is probably rated for continuous service, but the actuated device is usually not. A starter motor is a good case in point.

A schematic symbol for a push button momentary is shown in Figure 11.7.

11.6 Isolation/Step–Down Transformers

In a metal boat, an isolation transformer is a must. Even in other boats, a combination isolation and step–down transformer protects against electrolysis, and provides 115 VAC from 220 VAC circuits. Current between ship's ground, and the green AC wire was recently measured at a local marina. Three different boats were tested at widely separated docks. Current varied from about 170 milliamperes, to 250 mA. Obviously, anyone connecting to the green wire would supply their zincs to the marina.

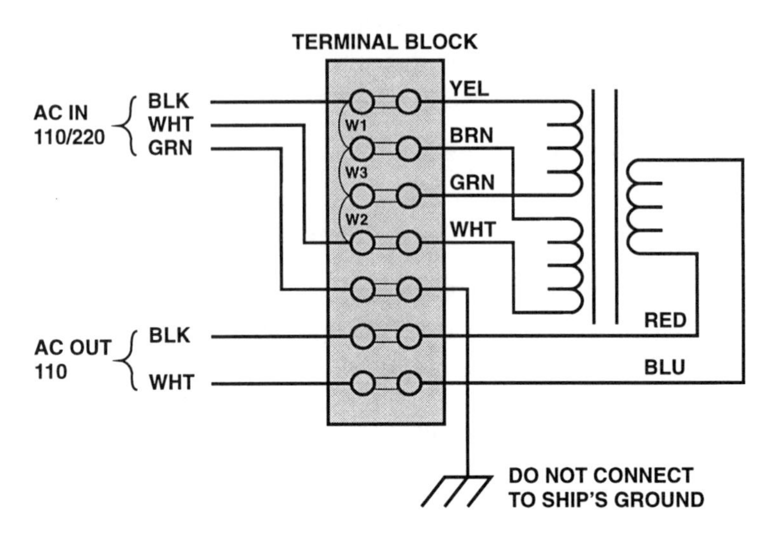

Figure 11.8: Isolation Transformer

One thing leading to increased neutral currents are the harmonics that personal computers, and televisions generate in their switching power supplies. With the situation worsening, an isolation transformer is a good investment.

In Figure 11.8 is the schematic for an isolation/step–down transformer. With the primaries connected in parallel, the transformer provides isolation, and a 1:1 ratio of input voltage to output voltage. With the primaries connected in series, the transformer not only provides isolation, but steps down the input voltage by half.

11.7 Transfer Relays

Relays are used to switch circuits that have too much current, or too high a voltage for a switch to handle directly. A relay is made up of an electromagnet that pulls a set of contacts together to complete a circuit. Actuation of the electromagnet may also interrupt a circuit.

Relays are made in several configurations and various contact materials for high voltage or high current applications. Figure 11.9, shows the

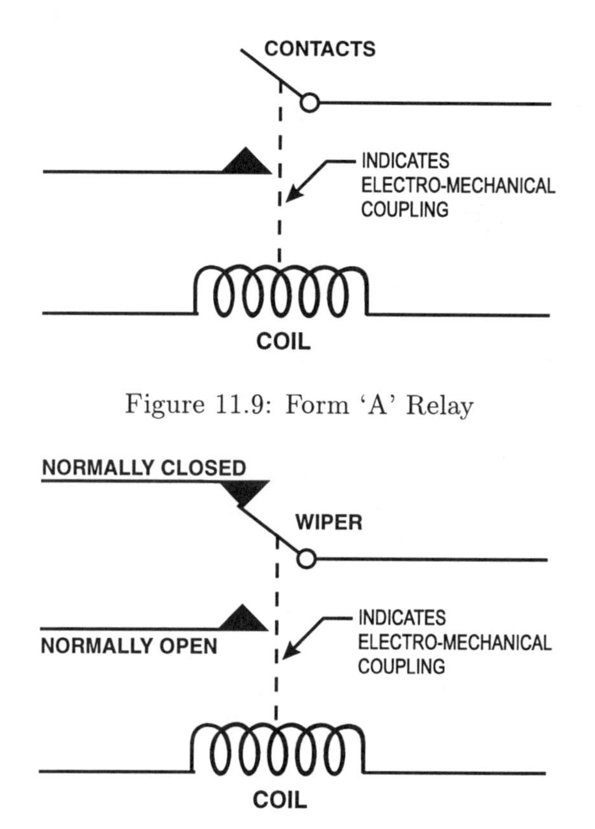

Figure 11.9: Form 'A' Relay

Figure 11.10: Form 'C' Relay

simplest of relays. It is two contacts that close when power is applied to the coil. Such a contact arrangement is called a *form 'A'* type. If the contacts are normally closed, and they open on application of coil power, the relay is said to be *form 'B'*.

Figure 11.10 shows a *form 'C'* relay. This type of relay has a moveable contact (or wiper) that swings between two stationary contacts. The wiper rests on the normally closed contact, and moves to the normally open contact when coil power is applied. It is conventional to draw relays in their unactuated state, but we learned years ago not to trust convention.

For the most part, you will see only form A, and form C relays. The starter solenoid is an example of a form A relay. Form A relays are used in high power spotlight systems, and sometimes used to turn pumps on, or battery chargers off. Form C relays are frequently used to switch power circuits from a generator source to an inverter source.

11.8 Solenoids

Solenoids are just large relays, usually in the Form A configuration. Be aware of contact ratings and coil ratings. Contacts are rated for maximum inrush current as well as continuous current.

Coils, the source of the magnetic field that closes the contacts are rated for intermittent or continuous duty. For instance, starter solenoids are only rated for intermittent duty. If you try to use a starter solenoid as a battery parallel switch, the coild will overheat and burn up after a few minutes of operation. The coil needs to be continuously rated if the solenoid is to be operated for more than 1–2 minutes.

Latching solenoids are also available. With a latching solenoid, two coils are provided. One coil is actuated to close and latch the contacts, while the second coil is used to unlatch and open the contacts. Coil current is only necessary to open or close the contacts and not during the time contacts are closed.

Some latching solenoids use a single coil. These require that current be supplied to the coil in one direction to close the solenoid, and in the other direction to open it.

11.9 Electrical Panels

Several organizations make standard electrical panels filled with circuit breakers. Each boat is different, so it is unlikely that a standard panel will exactly suit your needs. When you wish to incorporate controls for equipment such as battery chargers, or diesel furnaces, then a custom panel is the route to take. With a custom panel, it is also easy to integrate instrumentation.

A custom panel is not necessarily expensive, compared to a collage of standard units with the special functions thrown in on the side. A custom panel can also be arranged to take less space than the standard array, even if all circuits are controlled with breakers. Additional space can be saved by using fuse switches instead of breakers.

11.10 AC Shore Power Connections

There are several different outlet connections used at marinas in the U.S. Even in one locale, it isn't uncommon to find different connectors. With ample DC power aboard, we have never connected to the dock when staying temporarily at guest facilities. We see others that seem to connect to shore power before they have all the dock lines in place. They often use a *spider* that has various dockside connectors which are all tied to a single connector for their power cord. Spiders are sold at marine chandleries.

11.11 AC Outlets

Standard household outlets are generally sufficient for use aboard. Even the galvanized outlet boxes provide reasonable service. They can be covered with a marine stainless steel plate. Stainless AC outlets can be purchased at marine chandleries. If you have a damp location, we'd advise not putting any AC outlet there, but if you decide that you really need an outlet there, use a stainless one.

11.12 DC Outlets

The connector originally designed as a receptacle for a cigarette lighter has become a standard of some sort. It serves the purpose as a connector, but takes alot of room in the process. We have a small portable TV that came with a cigarette lighter connector, and our laptop computer uses one. The cigarette lighter connector is not applicable to topside service, however.

Unfortunately, there is not another 12 Volt connector that is a universal standard. There are connectors available in marine chandleries, but just about every outlet has a different brand. Many of the connectors are of questionable quality. We presently use a connector by Hella[1]. It is used for connection to spotlights and also for solar panels. It has a cap

[1]Hella Marine, Peachtree, GA.

when nothing is connected to it, so is protected under those conditions. The wire end of the connector is not sealed as well as we would like, but until a clearly superior alternative comes along, it will be our standard.

11.13 Alternator Bypass Units

Alternator bypass units are a mechanism that allows manual control of the alternator. In its simplest form, it is just a large rheostat connected between the alternator field and the battery. The rheostat is usually 4–6 Ohms and rated for 25 Watts or more.

Manual control of the alternator is better than waiting for the conventional regulator to charge batteries, but don't forget to adjust it down as the batteries approach a full charge ... alot of fried batteries and alternators have resulted from manual bypass units.

An additional level of sophistication involves a voltage sensing circuit driving a semiconductor. The semiconductor passes field current until the batteries reach a predetermined voltage. The sensing circuit then stops semiconductor current, allowing the conventional regulator to take over. The semiconductor in such controllers is used in a linear mode that requires it to dissipate considerable power. That makes it failure prone. The voltage drop across the semiconductor is also substantial, limiting the maximum current that might otherwise be obtained from the alternator. The devices are not suggested for alternators above about 70 Amps.

11.14 Alternator Regulators

Alternator regulators are solid state devices that sense voltage, and adjust field current to maintain the voltage at its sense point. The typical alternator regulator was designed for use in the automobile. The starting battery in a car is not normally discharged much, so a regulator to provide a fast and safe recharge is not required. The automotive regulator is not suited to charge deep cycle batteries. It doesn't provide a fast charge, but will eventually go on to overcharge batteries if the engine is operated for long hours.

Figure 11.14 shows a regulator made by Ample Technology[2]. The

[2] Ample Technology, Seattle, Washington

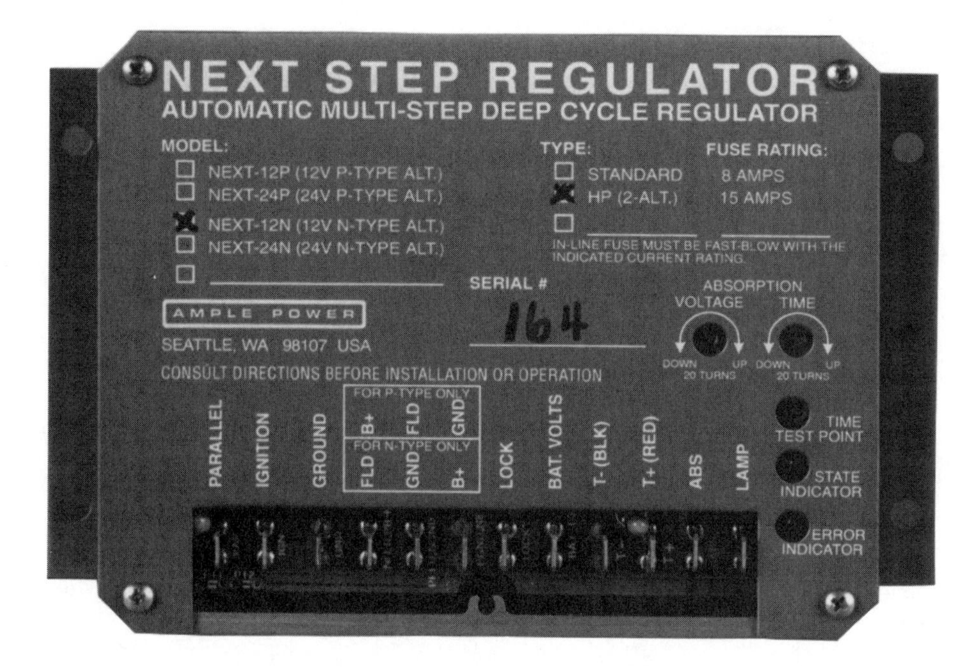

Figure 11.11: Next Step Deep Cycle Regulator

alternator regulator accomplishes a fast full charge, using an appropriate higher absorption voltage. The regulator trips to a lower float voltage after the absorption charge, thus overcharges are avoided. The regulator operates in conjunction with a temperature sensor that is placed on the battery. If battery temperature begins to rise, then the regulator cuts back the applied voltage. The sensors are not simply on/off, but operate linearly to constantly track temperature changes.

11.15 Battery Isolators

Battery isolators are diodes used in series with charge sources to charge multiple batteries. Like any other apparatus, there are pros and cons regarding their use. We've seen enough problems created by isolator usage that we can no longer recommend them for most systems, however, the isolator more or less solves a problem that many people are not willing to deal with ... choosing which battery to charge. An isolator also provides battery protection in case of alternator failure.

Isolator problems are still the same, unequal voltage drop in the diodes, and increased power required from the alternator. An unsensed battery will charge to the peak voltage of the ripple present on alternator outputs. This usually means that the starter battery is overcharged.

If you elect to use an isolator, be sure that you are using a regulator that senses the voltage at the batteries. Otherwise, you will be severely undercharging your batteries due to the voltage lost in the isolator.

When selecting an isolator, be certain the unit is rated for the maximum current that the alternator can produce when it is cold.

Isolators should be mounted with their heatsink fins vertical for best heat conduction to the atmosphere. From a wiring point of view, it pays to mount the isolator close to the batteries, but don't put it in a battery compartment that is not well ventilated ...isolators do generate heat.

If you use an isolator that has more output posts than required, simply leave the unneeded terminals open. Don't short two outputs together since this will aggravate the difference in voltage drop between diode legs.

11.16 Charge/Access Diodes

A Charge/Access Diode is a device that allows a bilge pump or electronics to be operated from two batteries. Sensitive electronics can be protected from low voltage transients that occur when an engine is started. The battery which is not used for starting supplies power to the electronics.

A bilge pump that is powered through a Charge/Access Diode has the capacity of both batteries available, rather than just one of them.

11.17 Ground Fault Interrupters

A ground fault interrupter, GFI, is a device that interrupts the AC circuit whenever current to safety or earth ground is detected. The detector operates by sensing the AC current in both the hot and neutral sides. If the two currents are not equal, then a path to ground must be present. If that path is yourself, then you'll be glad to have a GFI installed because it will save your life.

Ground fault interrupters are made in different ratings and cost accordingly. Most certainly, an AC outlet in the head should be equipped with a GFI. The galley is another good spot to have protection. Anywhere you have water, which can make a good connection to ground, consider a GFI.

Sometimes GFIs will trip when battery chargers are turned on. This is because the manufacturer of the charger is using an input filter to reduce noise, and one leg of the filter is tied to the safety ground. If you have a sensitive GFI and a the charger uses a big filter for effective noise reduction, then nuisance trips result. For this reason, GFIs are not recommended in the main power input to your boat or RV.

Chapter 12

Wire, Terminals and Aids

12.1 General Information

It's easy to think of wires as being all alike, and one terminal block the same as the next. Like anything else, there is a difference in quality and suitability between products. In this chapter, we describe some of the wire and wiring devices that go into an electrical system, and the specifications that assure a high quality and reliable finished product ... your electrical system.

Wire is measured according to the American Wire Gauge, AWG, that specifies the cross sectional area of any given wire. The higher the gauge, the smaller the wire, until the #0 gauge is reached. Then, the more zeroes used, the bigger the wire. That is #000 is bigger than #00. Zero gauge wire is sometimes pronounced as *awt* gauge, and #00 is called double awt.

12.2 Wire Types

Wires are made in solid copper strands, such as that used in houses, or it is available in bundles of smaller strands. The strands may be bare copper, or they may be tinned, that is the copper wires are covered with a tin–lead alloy. The purpose of the tinning is to prevent corrosion. Bare copper wires that have been corroded must be scraped to remove copper oxide which is an insulator.

12.2.1 Bare Copper Wire

Bare copper wire has a limited place on boats. It can be used to bond through–hull fittings, engine and stern tube together. If large enough, it can be used to connect chain plates to ground for lightning protection. The surface of the wire will quickly form a black copper oxide in the marine environment. As long as the wire is not directly exposed to salt-water, a large diameter copper wire will not present a corrosion problem if securely fastened with a hose clamp to the fitting. Vibration must be prevented by securely fastening the wire.

Solid copper wire typically used in homes is not a good choice for marine or RV use, even, or perhaps especially, in AC circuits. Terminations to solid wires are more prone to corrosion than stranded wires. The terminations are also more prone to fail under vibration. Vibration of solid wire can even fatigue the wire and ultimately cause it to break.

12.2.2 Stranded Wire

Stranded wire is much preferred over solid wire. Stranded wire is generally made up of sub–bundles of strands. Only the least expensive stranded wire is made up of all equal strands in a single bundle. Stranding for wires is listed as X/Y where X is the number of bundles, and Y is the number of strands in each bundle.

For the same wire gauge, some manufacturers use fewer bundles of larger diameter wire than others. The more the bundles, with finer strands, the greater is the cost of the wire. Wires with finer strands are also more flexible, so they are easier to work with. Conductivity is generally better as well, so for a top of the line wiring job, use wire with fine strands of wire. Table 12.1 show bundle and strands for some wires. Note that different strandings are available for any given wire gauge.

As mentioned, stranded wire comes in bare copper and pre–tinned. Pre–tined wire naturally costs more, but offers better protection against corrosion. It is always possible for water to enter a wire at the end, and it will wick itself into the wire for a long distance.

If you feel comfortable with your electrical system, and can troubleshoot it under emergency conditions, then you might consider untinned stranded wire above the floors where moisture is not so great. For the most reliable system, use pre–tinned stranded wires. In the overall cost

Wire Gauge (AWG)	Bundles/Strands
18	1/18
18	16/30
16	1/16
16	26/30
14	1/14
14	41/30
12	1/12
12	65/30
10	1/10
10	105/30
8	133/29
6	133/27
4	133/25
2	133/23
2	665/30
1	259/25
1	836/30
1/0	1045/30
2/0	1330/30
3/0	1661/30
4/0	2107/30

Table 12.1: Wire Stranding

Property	Insulation			
	Polyvinyl– chloride	Poly– ethylene	Poly– propylene	Teflon
Oil Resistance	E	E	E	O
Heat Resistance	G	G	E	O
Sun Resistance	G	E	E	O
Ozone Resistance	E	E	E	E
Abrasion Resistance	F	F	F	E
Flame Resistance	E	P	P	O
Solvent Resistance	P	P	P	E
Flexibility	P	G	P	O

Table 12.2: Properties of Plastic Insulations

of a boat, the difference between untinned and pre–tinned wire is not significant.

12.2.3 Wire Insulation

There are quite a few plastics that are used for insulation. Properties of some are given in Table 12.2. The letters under the insulations stand for **P**oor, **F**air, **G**ood, **E**xcellent and **O**utstanding.

The Underwriters Laboratory has many different specifications regarding wire insulations. An insulation quite suitable for marine use is covered by UL[1] specification #1015. To meet this specification, the insulation must be flame retardant, be rated at 105°C, and 600 Volts. Insulation is usually polyvinylchloride with flame retardant additives.

12.3 Crimp and Solder Terminals

Crimp and solder terminals are sold with and without insulation. Generally, you will want the type with insulated barrels. There are two types

[1]Underwriters Laboratories, Inc.

of insulation available. Nylon is tougher and stands up to heat and age better than vinyl. The price differential is about 30%.

Terminals are also made with an insulation that acts as a sealing compound. Once the terminal is crimped onto the wire, a heat gun is used to melt and activate the sealant.

Terminals are made as forked devices, or with a full ring. The latter type will not shake loose if the screw holding it loosens a little. Loose screws shouldn't happen, of course. If they do, then how long can it be before the screw falls completely out?

We don't hold the prejudice against forked terminals that is common. If the wire bundle is properly secured and wires are not stretched or stressed, then a forked terminal will stay in place even with a loose screw. Forked terminals are easier to connect to terminal blocks because it isn't necessary to completely remove the screw. Ring terminals are undoubtedly superior for sloppy wiring, and best when wiring is not expected to be changed.

12.4 Terminal Blocks

A large variety of terminal blocks are available. The most important characteristic of a terminal block is the type of hardware and screw that is used. Many terminals are made of steel which quickly rusts. When you go to buy terminal blocks, carry along a magnet to test for steel hardware.

The terminal blocks suited for marine applications utilize brass hardware that is chrome plated.

Terminal blocks must be sized according to the amount of current that they conduct, and the voltage that is present on the terminals. The voltage governs the spacing between terminals. Be sure that the blocks you use are rated for AC service, particularly for 220 VAC. Figure 12.3 shows the proper size terminal for several current ratings.

12.5 Distribution Terminals

A bolt on the engine housing is often used as a negative distribution terminal, while the center position on the battery selector switch is often

Screw Size	Current Rating
#5	15
#6	20
#8	30
#10	50
#12	70
#14	90

Table 12.3: Terminal Block Size versus Current Rating

the positive distribution point. Neither of these is meant to be. For a few dollars, positive and negative distribution posts can be installed. Not only do they make wiring easier, but the integrity of the system is higher.

Wires under engine bolts are often left dangling after engine service because the mechanic forgets about the electrical functions. Using the engine block as a conductor is not a good idea, especially with high output alternators.

12.6 Wire Fastening

Wires which are not securely fastened to a rigid surface are prone to chafe. Loose wires are also easier to inadvertently snag, leading to open circuits. Many different types of fasteners are available which can eliminate problems with loose wires.

12.6.1 Wire Anchors

Wire anchors are devices that can be screwed to a bulkhead or beam, and then used as an attachment point for lacing cord or nylon ties. Figure 12.1 shows two different types. The one on the left takes a single screw, so is quick to mount. The nylon tie goes under one edge of the anchor. This type of anchor is sufficient for much wiring, as long as the open end on the anchor is oriented so that the nylon tie is not under strain away from the end.

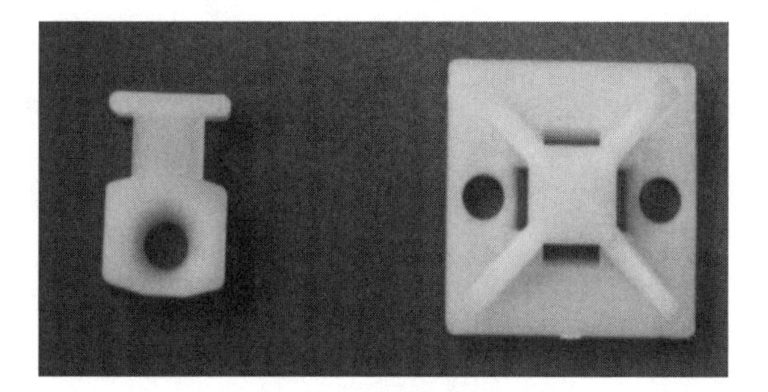

Figure 12.1: Wire Anchors

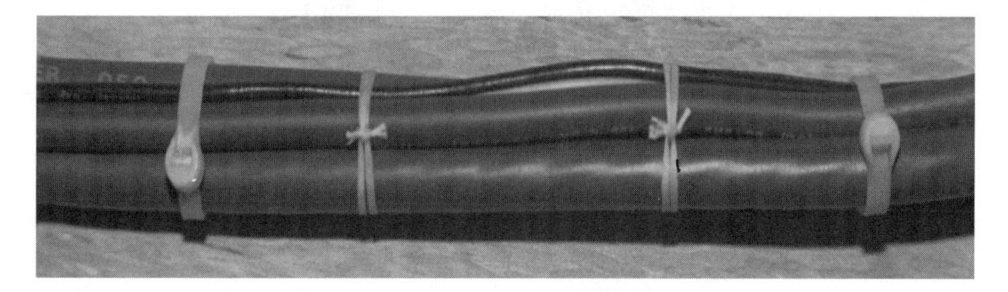

Figure 12.2: Lacing Cord

The anchor on the right of Figure 12.1 mounts with two screws. The nylon tie goes through the center of the anchor and is thus fully secured against strain in any direction. This type of anchor is preferred for heavy wiring bundles, or bundles under strains from different directions.

12.6.2 Waxed Lacing Cord

Before the advent of nylon ties, waxed lacing cord was used to secure wires in a bundle. We learned to tie a clove hitch around wires long before we knew what the knot was called. Lacing cord comes in either black or white, and in different strength ratings.

Waxed lacing cord has a few advantages over nylon ties. First, it is less expensive, and one size fits all since you cut the cord from its roll after making the tie. The cord is less environmentally dangerous, and it doesn't leave scratches on hands and arms when brushed against, as will a nylon tie that has been cut off. Except for threading through wire anchors, lacing cord can be just as quickly applied. We prefer the look of lacing cord between anchor points. Figure 12.2 shows a wire bundle that is laced between anchors with a black cord.

12.6.3 Nylon Ties

Nylon ties come in various widths and lengths, as well as in black and white. You can even get nylon ties with a screw hole so that no anchor is required. We generally avoid this type of tie, since you have to remove the screw and re–attach it to change the tie. The angle on the tie never lines up with the mounting plane for the screw, so the ties are often difficult to get mounted.

Black ties last longer in ultraviolet light, so always use them outside. Even inside, the white ties yellow quickly with age, giving an old look to the wiring runs. Naturally, black ties cost more.

12.6.4 Wire Raceways

Some people like the appearance of wire raceways. With their covers on, and wires neatly exiting the raceway as necessary, the appearance is indeed pleasing. We don't often see too many such systems. More often, the raceway is overfull with additional wires running next to it. The cover has long since been discarded, so the raceway only serves to use up space in an overcrowded wiring channel.

This situation is easy to understand. First, the builder doesn't want to spend money needlessly on a large raceway. One is selected which meets the builders wiring plan with a small margin for later additions. Later additions turn out to be extensive. When a real power system is added, wire sizes must be bigger, and often more of them. Electronic devices are often installed after yacht delivery, and the extra margin allowed by the builder is soon gone.

No installer wants to be the first to go outside the raceway. Sometimes this leads to the installation of smaller wires than are appropriate. Wires are often crammed into too little space, eliminating any hope of tracing them later during troubleshooting. If you are convinced that a raceway is justified, make it generously oversized, and prepare to have a few wires running outside.

12.7 Wire Protection

12.7.1 Tape and Heat Shrink Tubing

PVC tape is often used cosmetically to cover poorly terminated ends. Its insulation properties are good, if used in several thicknesses. If applied in warm weather, it does a fair job of sealing.

For the best appearance, and the best protection against moisture, use heat shrink tubing. Heat shrink tubing is now available with a silicone interior. When it's heated and shrunk, the silicone activates to make a superior barrier. If you use untinned wire, heat shrink tubing is good insurance.

12.7.2 Protective Chemicals

One of the best protective chemicals made is petroleum jelly. You may have some around the house to use for cuts and abrasions. For corroded battery terminals, first wash the terminals with a solution of baking soda and water. After the terminals are thoroughly dried, rub on a thin layer of petroleum jelly. Jelly can also be applied to terminal blocks and the terminals themselves. For terminals under the floors where humidity is often high, petroleum jelly will protect as well, if not better, than many other expensive chemicals. Butt splices beneath the floors should have jelly pushed into them after the crimps are made, and then protected with heatshrink tubing.

There are also paint–on and spray–on chemicals. An excellent paint or dip silicone for protection of printed circuit assemblies is SR900 made by General Electric[2]. Be careful not to get any of the sealant in connectors or terminals, since it is an insulator. It is also a hazardous material to use, so don't get any on your skin, and always dispose of waste appropriately.

Any electronic outlet carries various spray–on chemicals for cleaning and protecting electronic circuits. We don't know much about the performance of any of them, because the silicone mentioned above, and the liberal use of petroleum jelly has always given us the protection expected.

[2]General Electric, Waterford, NY.

Chapter 13

Tools

13.1 General Information

Having the right tool for the job usually means the difference between mediocre workmanship and a job well done. Tools always appear expensive, but are seldom so. Instead, tools are an investment, particularly high quality tools that may offer a lifetime of service.

What tools are required to wire a boat? How are such tools used? Where can those tools be purchased? These are some of the questions that are answered in this chapter.

13.2 Wire and Cable Cutters

Figure 13.1 show three wire and cable cutters. The small cutters on the left can be used for small wire ... 8–22 gauge. The center cutters are useful for wire from #2 and smaller. On the right is shown a cable cutter. It will cut wires as big as #0000.

The smaller wire cutters are readily available at any electronics outlet. Some electrical suppliers also carry them. Beware of poor quality cutters that quickly lose their edge, and then fail to make a clean through–cut. We have a wire cutter made by Xcelite[1] that continues to work well after many years of use.

[1]Cooper Industries, Xcelite Brand, Apex, North Carolina.

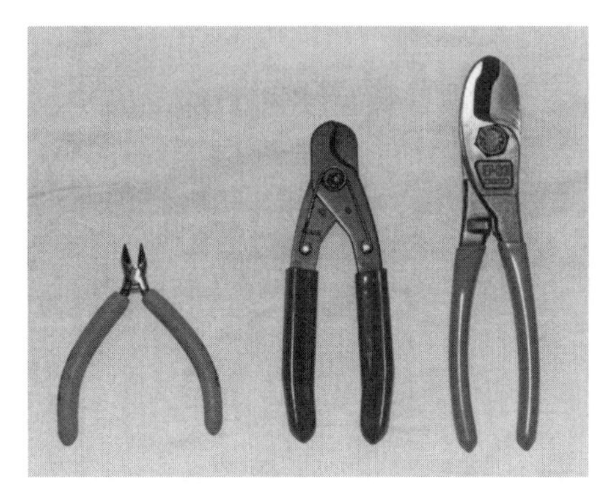

Figure 13.1: Wire Cutters

If you are only going to cut a few large wires, say #00 gauge for battery wiring, it is possible to do so with the larger wire cutters shown in the center of Figure 13.1. You will have to nibble away at the wire until it is completely cut. The problem with nibbling through the wire is the ragged end ... it can be a real nuisance to slide a terminal lug over the uneven strands.

For a professional cut, and one that accepts terminals easily, you will need a cable cutter such as that shown on the right in Figure 13.1. It is relatively inexpensive, yet makes a clean cut through large cable. The cutters can be purchased at some electrical suppliers.

The hacksaw has been used to cut large battery cable. It leaves very jagged edges, making it extremely difficult to get a lug over the wire. When there is no other way, however, the hacksaw does work.

13.3 Wire Strippers

For years we used small wire cutters as wire strippers. By carefully applying the right amount of pressure, the insulation could be cut and stripped, without harming any of the wire strands. Insulations have grown tougher over the years, particularly the high quality insulations that should be used in marine applications. Now we only use wire strippers for stripping wire.

Figure 13.2 show three types of wire strippers. The stripper on the

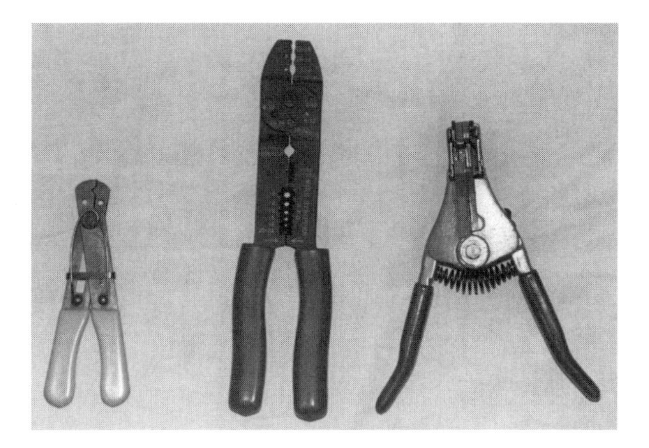

Figure 13.2: Wire Strippers

left is an inexpensive unit. It has an adjustment screw on one arm that sets the jaw opening for different gauges of wire. There are no wire gauge calibration marks on the adjustment screw so adjustment is strictly a trial and error affair. With adjustment difficult, you will either need a stripper for each gauge of wire in use, or you will need to leave the adjustment for the smallest wire in use, and carefully strip larger wires so as to not cut strands. This latter technique takes much practice ... its easier than using wire cutters as strippers, and works well on all except some of the military specified insulations.

The next level of wire strippers are those that are part of wire crimping tools. Several different cutter gauges are fabricated in the handle. This type of stripper is illustrated by the center part of Figure 13.2. On the cheaper combination crimpers and strippers, the strippers quickly lose their edge. Insulation is then sort of *stretched* away from the wire.

For a lot of wire stripping, with a minimum of practice, the best strippers are those shown in the right of Figure 13.2. These units are designed to hold the wire as well as strip the insulation. This type of stripper covers a limited range of wire sizes, say #14–22 gauge. Another unit is necessary for wires #8–12.

For large wire and battery cable, dedicated strippers are probably too expensive for the amateur. Using wire cutters or the inexpensive strippers shown on the left in Figure 13.2 may be used, as can the ordinary pocket

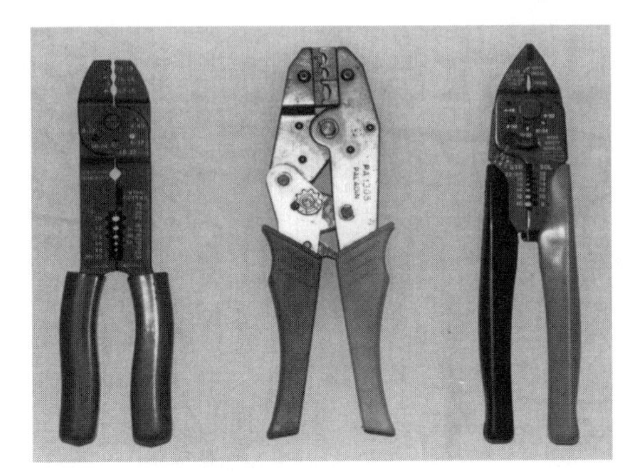

Figure 13.3: Wire Terminal Crimpers

knife.

Using a knife to strip wire without damaging the strands takes considerable practice, particularly on finely stranded wire. To avoid cutting stranded wire, it is best not to slide the knife blade over the wire. Instead, *roll* the blade while pressing firmly against the insulation. You may prefer to pinch the wire between your thumb and the blade, and gently roll the wire over the blade. Do a small section at a time until you have rolled the wire completely and the insulation is cut through to wire.

After cutting the insulation, you may still have difficulty pulling the severed insulation from the wire. On tough insulations you may have to make a longitudinal slit in the insulation. Again, avoid sliding the knife blade, although it isn't as apt to cut wires longitudinally.

13.4 Wire Crimping Tools

Figure 13.3 show three crimping tools. On the left is shown a combination crimper/wire stripper that is least expensive. In fact, some of the cheap crimpers are now made so poorly that they are not good for crimping because of flex in the ends. You will probably want to use the combination crimpers for all small terminals due to their low cost. Choose carefully. Find a unit with thick material, and comfortable grips. We have an aging unit from Sears[2] that works well. Recently we bought another pair for a second tool box. Made overseas, the unit actually flexes enough to cut through the terminal when force is applied to the terminal. Even when it doesn't cut the terminal, it damages the insulation. Of what value is such a tool? We returned it, of course.

In the center of Figure 13.3 is a calibrated crimper with a racheting action. To make a crimp requires enough force to carry the rachet action through to its finish. The force necessary to achieve full rachet pressure is adjustable, and once adjusted, every crimp is made with the same pressure. While these tools are expensive ($40), they are essential when a large number of crimps are to be made, and the quality of the crimp connection must be high. The crimper shown is made by Paladin[3].

On the right of Figure 13.3 is a quality crimper that isn't calibrated, but does have marks to show how far to squeeze the handles together. This tool costs more than the crimpers on the left, but it is less than the calibrated crimpers in the center. The tool includes a cutter, wire stripping slots for #22–10 gauge, and bolt cutters. For occasional use, this crimp tool is the best value. The manufacturer is Thomas & Betts[4], Model WT–2000.

All crimps should be tested for mechanical strength. A crimped connection on small wire should be able to sustain at least a 5 pound pull. If you are making crimp connections that will not withstand the 5 pound pull test, expect some of them to come loose later.

13.5 Battery Cable Crimping Tools

The crimper shown in Figure 13.4 is a unit that crimps terminal to large wire. It comes with a set of dies that are swapped in the jaws for different wire gauges. You probably won't want to purchase these crimpers unless you have a large job to do since they cost about $300. They are made also by Thomas & Betts.

---

[2]Sears Roebuck Company, Chicago, Illinois.
[3]Paladin, Newbury Park, California.
[4]Thomas & Betts, Raritan, New Jersey.

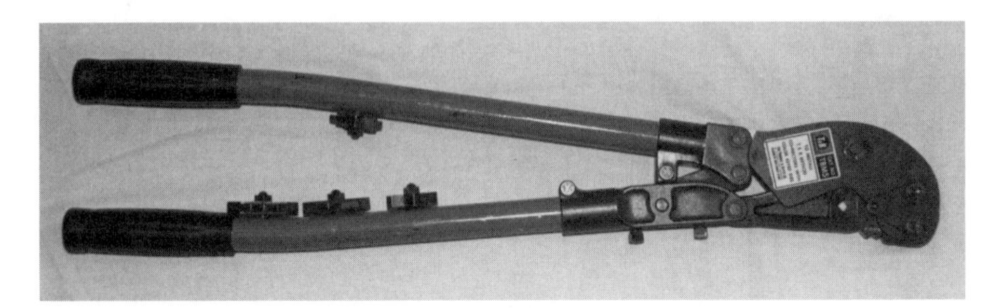

Figure 13.4: Battery Cable Crimpers

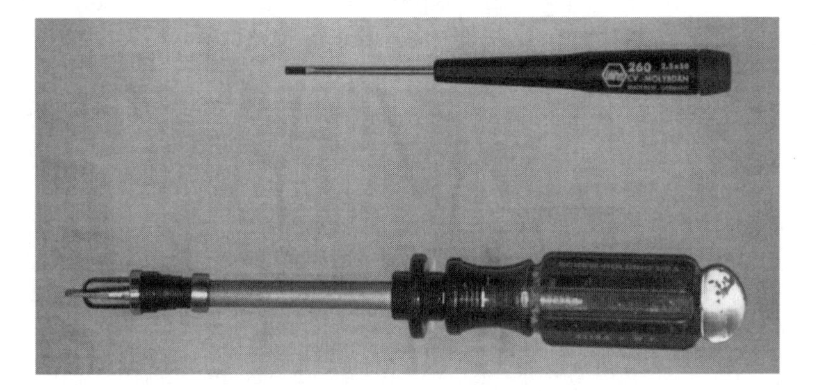

Figure 13.5: Screwdrivers

13.6 Awl

An awl may not seem much like a wiring tool. In fact, our biggest use of an awl when wiring, is to punch a starter hole for a screw which fastens a wiring anchor. An awl can be used on all soft woods, but hard woods often require a small hole be drilled. This is especially required for thin or narrow pieces of hardwood that may crack if a proper hole is not drilled before screwing on the tie anchor.

13.7 Screwdrivers

All small screwdrivers are applicable for wiring chores. There are two speciality screwdrivers that are very useful. The first of these is shown

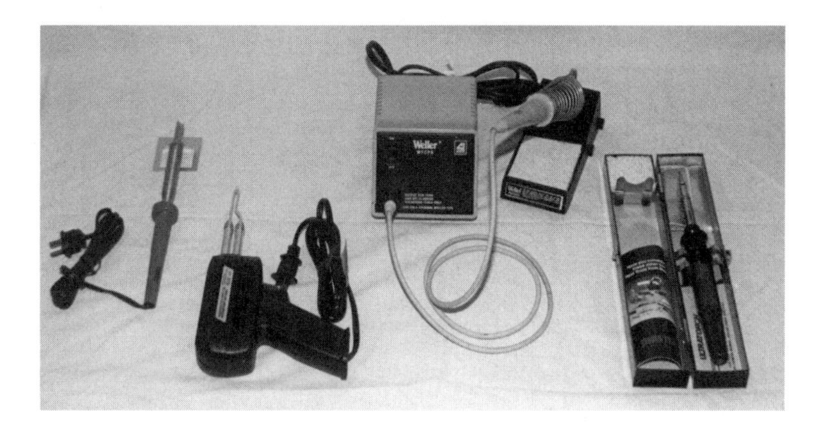

Figure 13.6: Soldering Tools

on the bottom of Figure 13.5. It has some *jaws* that clamp a screw to the screwdriver. This tool is most useful when fastening ring lugs onto a terminal block. Without the jaws, fighting the screw through the ring lugs and into the terminal block runs a high risk of losing the screw.

The second speciality screwdriver you'll want is often called a jeweler's screwdriver. The small screwdriver on the top of Figure 13.5 is used for the small screws in printed circuit terminal blocks. It is also used for potentiometer adjustment.

13.8 Soldering Tools

Figure 13.6 shows four types of soldering tools. The soldering iron on the left is a high capacity unit suitable for soldering larger wires, say #10 to #4. The heat of the iron is not regulated, so you must be careful not to overheat the work which can burn the insulation. The soldering iron is being heated whenever it is plugged into the AC circuit.

The second iron from the left in Figure 13.6 is a trigger operated soldering iron. That is, the iron is being heated only when the trigger is activated. Such an iron is generally not as easy as the iron shown on the left, because each time you use it, you must wait for a short period while the iron warms up. For a big electric iron, we favor the left most tool.

The third tool from the left in Figure 13.6 is a thermostatically controlled soldering iron. It finds principal use as an electronic assembly and

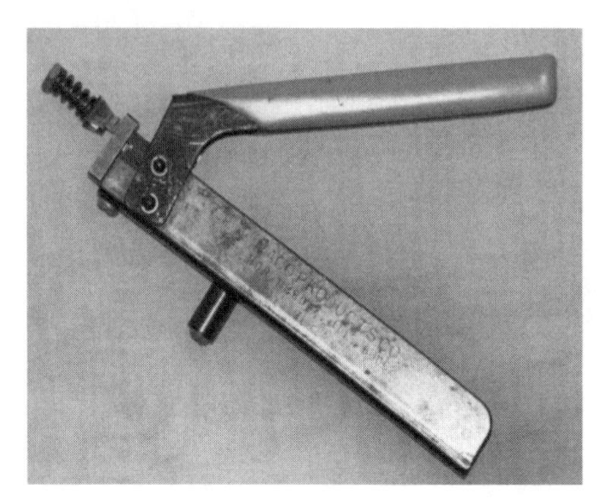

Figure 13.7: Nibblers

repair tool. If you are into soldering all your small terminals after they have been crimped, the regulated soldering iron works the best. For this, you will probably want the small pointed tip that makes good thermal contact to both the wire end and the terminal itself.

On the far right of Figure 13.6 is a butane heated soldering iron. The particular unit shown also has a torch adapter. The unit has an adjustment that controls heat delivered to the tip. With the small soldering tip, wires are easily soldered to terminals. By replacing the soldering tip with the torch, large terminals can be soldered to the battery cable. Butane soldering irons are indispensible. You don't need electricity to operate them, which makes them perfect for wiring work aloft. On the other hand, they do have an open flame, so we wouldn't use one on a gasoline powered boat.

13.9 Nibblers

Nibblers, such as the unit shown in Figure 13.7 are valuable for cutting holes in thin aluminum, copper or brass. They will even work on very thin steel. Nibblers operate by cutting away a small chunk of material each time the handles are fully squeezed. To use a nibbler to make a rectangular hole, first drill a hole inside the rectangle that is to be removed. Now insert the nibbler cutting edge through the hole and start nibbling around the perimeter of the rectangle. It is best to leave some excess material which can be removed later with a file, smoothing up any ragged edges.

13.10 Drill Motors and Hole Saws

It is unlikely that you can accomplish much wiring without the aid of a drill motor. A small 1/4 to 3/8 HP unit is best for all around use, but a larger one is often necessary to use larger hole saws. Hole saws, of course, are the principal way to open access through a bulkhead.

Drill motors are also useful to twist wires, as explained elsewhere in the book.

13.11 Multimeters

Digital multimeters are available from Radio Shack[5], and any electronic supply. There are a few manufacturers of quality meters. We have found that Fluke[6] meters are accurate, impact resistant, and run for long periods on their batteries.

When purchasing a digital meter, you want to be able to measure both AC and DC voltage, AC and DC current, resistance, and diode conduction. It is handy if the meter has a beeper that sounds when a diode conducts. The diode function can be used for continuity checking and the beeper means that you won't have to look away from the work to know if there is continuity.

13.12 Thermometer

No tool box is complete without a thermometer of some sort. Refrigeration thermometers are useful for more than just measuring refrigeration temperatures. Batteries are temperature dependent, and some regulators[7] sense temperature and adjust their regulation accordingly. A thermometer is not required to initially adjust the regulator, however it is a useful tool to check on temperature tracking.

[5]Tandy Corporation, Fort Worth, Texas.

[6]John Fluke Manufacturing Company, Everett, Washington.

[7]Next Step Deep Cycle Regulator, Ample Technology.

Operating temperatures of alternators, isolators, chargers, and inverters are important. Testing and recording under different load conditions can prove useful later when trouble is suspected.

If you have a digital meter, consider a temperature probe adapter. We use an adapter made by Fluke that covers a wide range of -58 to 1832°F. The probe is not actually rated for submersion, but we have used it to measure fresh water temperature for an engine.

13.13 Pocket Compass

Testing for high currents can be done with a pocket compass. It isn't very sophisticated, but it works. If you can't afford a clip–on ammeter to go with your digital meter, buy a cheap compass. As the compass is placed next to wires carrying current, the needle will deflect.

13.14 AM Radios

AM radios make a good tool to sniff out sources of noise. By tuning the radio at the lower end of its band, and not on a radio station, you can pick up electrical noise through the radio antenna. The closer you get to the source of noise, the louder the radio will sound.

You may need to shield all of the radio except its antenna coil with tin foil, and ground the foil. Radios that work best for noise sniffing are ones with long antenna coils.

13.15 Alligator Clip Leads

Alligator clip leads are wires with clips one each end that can be attached quickly to make temporary circuits. The clips have an opening jaw that looks quite like the jaw of an alligator. Though not a usual part of an electrical system, alligator clips can be used in place of a switch or fuse to make a circuit. They are often useful during testing of various parts. Whenever a wire is needed that you don't wish to make permanent, consider an alligator clip lead. Use them to temporarily jumper past defective connections.

Clip leads have ratings, like everything else. For temporary operation of pumps and blowers, clip leads of #14 gauge are appropriate. Carry a couple of them ...sooner or later, you'll find a handy use.

Chapter 14

Schematics

14.1 General Information

This chapter contains schematics for various electrical circuits. We have attempted to present the schematics so that no special knowledge is required.

It should be noted that we use the ground symbol for all wires that return to negative distribution, rather than showing the actual ground wire.

14.2 Grounding

Proper grounding is an important subject that we covered in greater detail in *Living on 12 Volts with Ample Power*. All ground wires for instrumentation and electronics should be connected at a central distribution post, or ground buss. Some non-sensitive devices such as pumps and fans can share a common return to the negative distribution terminal. Be careful sharing ground return wires for lights with pumps ... your lights may blink on every stroke of the pump.

Instruments that take measurement seriously will have a power return gound and a measurement ground. Those two circuits eventually get connected, and where that connection is, becomes the *reference ground* point. All measurements will be referred to this point, which typically will be the negative distribution terminal.

Figure 14.1: Distribution Bars and Post

Most boats collect ground wires on the engine block. This is not a good place to collect wires, if for no other reason than vibration. Wires should be collected at a terminal of some kind. Not only will this provide an electrically secure ground, but it also aids in expansion wiring or troubleshooting. Figure 14.1 shows a couple of distribution bars and a distribution post. These can be used as either postive or negative dsitribution points.

14.3 Bilge Pumps

Figure 14.2 shows two bilge pumps, two float switches, and an alarm module. The bilge pumps can be controlled manually by switches on the electrical panel, or by the float switches in the bilge. The float switches can be disabled by key switches. It is intended that the key be removable only with the float switch enabled. Power for the bilge pumps is derived from either of the battery banks via Charge/Access Diodes.

An alarm module and audible alarm are also shown. The alarm module, called provides two types of alarm signals to the audible alarm. Whenever bilge pump #1 starts, the alarm *chirps*, that is, it beeps for about a second. Whenever pump #2 is activated, the alarm sounds steadily.

It is intended that pump #1 is the primary pump, and thus is mounted lowest in the bilge. Pump #2 is a standby pump which only operates if pump #1 fails to keep the bilge dry. Pump #2 is mounted higher in the bilge (or at least its float switch is). The chirp that occurs on start up of

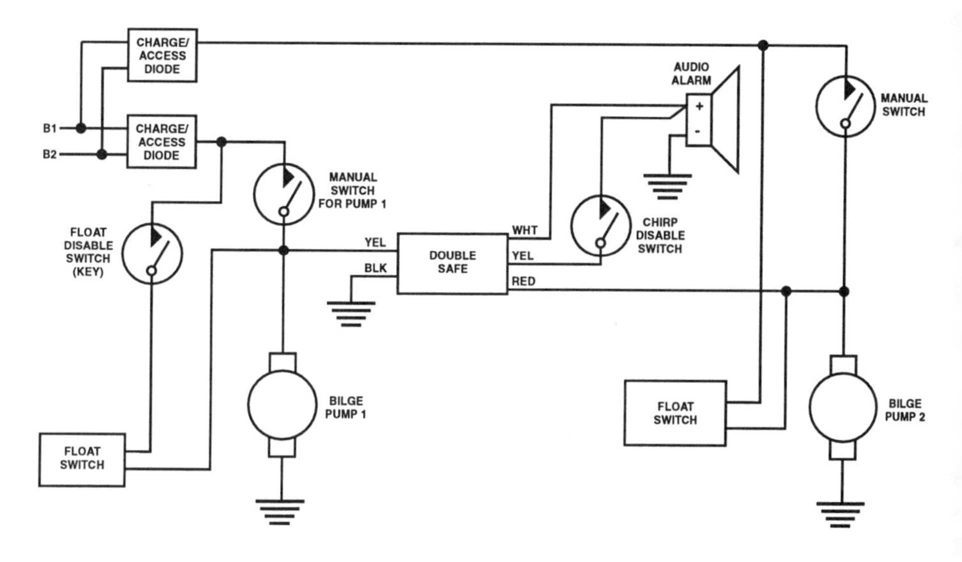

Figure 14.2: Dual Bilge Pump with Alarm

pump #1 can be disabled as shown with a toggle switch when it might be objectionable.

In a boat that leaks, the period between beeps should be constant unless the leak increases. A high rate of chirps means that the bilge pump is working often ... perhaps too often?

Key switches are used to disable the float switches. Normally, it is good practice to let the pumps work whenever the float switch detects high water, however, when under way in turbulent conditions, the remaining water in the bilge can slosh enough to keep cycling the pump on for a few seconds. Intermittent tripping of the float switch can be a serious loss of battery power over the course of a day. Chirps from the Double Safe module will indicate such a condition. If you do disable the float switch, be sure to have the watch regularly check the bilge and pump out manually.

14.4 Pressure Water Pumps

Figure 14.3 shows a pressure water pump. The pump is enabled by a switch on the control panel. Whenever low pressure is detected, the

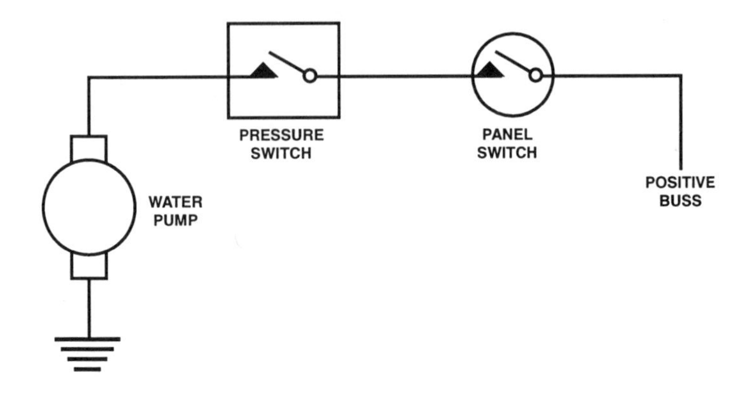

Figure 14.3: Pressure Water System

pressure switch closes, activating the pump motor.

14.5 Electric Windlass

Figure 14.4 shows the wiring for an electric windlass. The windlass is enabled by a switch on the control panel which feeds power to a foot switch. The foot switch does not directly drive the windlass, since it is not rated for the current drawn by the windlass. Instead, the foot switch activates a solenoid. The solenoid, in turn, activates the windlass motor.

A diode is shown across the coil of the solenoid. Its purpose is to prevent arcing across the foot switch whenever it is released. The diode can extend the life of the foot switch significantly. It can also prevent arcing inside the solenoid coil which can happen as moisture permeates the coil and breaks down the insulation. The diode is a 1 Amp silicon unit available at any electronic supply.

The solenoid contacts must be rated for the current draw of the windlass motor. Starter motor solenoids can be used. Sometimes high current solenoids can be found at surplus electronic outlets. We found a couple of hermetically sealed, military surplus solenoids rated for 400 Amps when we were building our ketch, and only paid $9.00 for the pair.

A diode is also placed across the windlass motor. This will stop motor coasting, and also protect the solenoid contacts from arcing. An appropriate diode for this application is the snubber sold by Ample Technology.

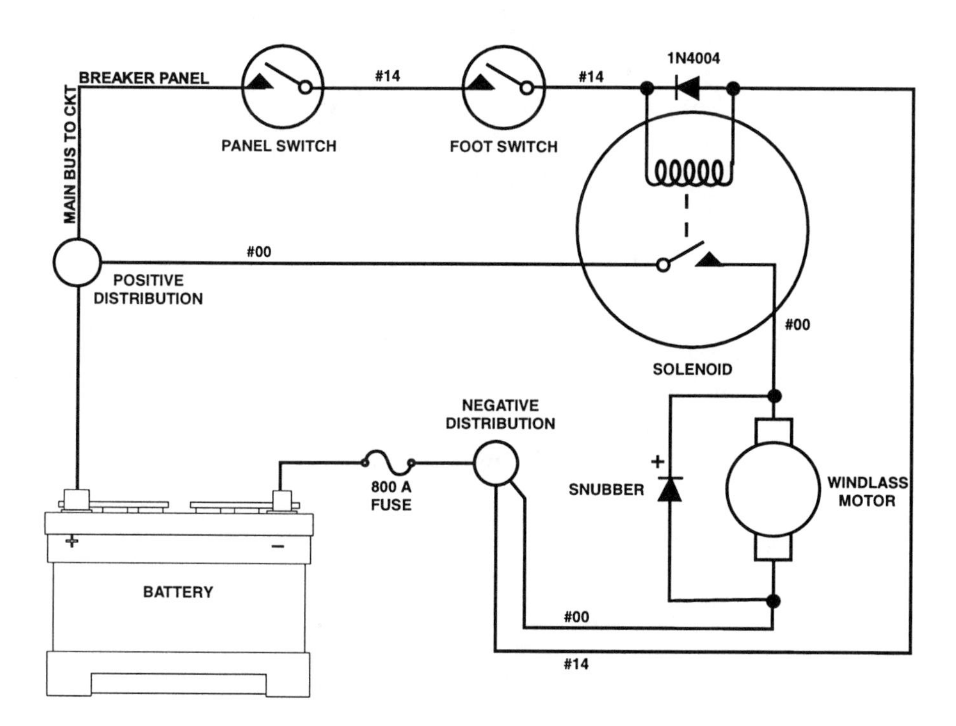

Figure 14.4: Electric Windlass

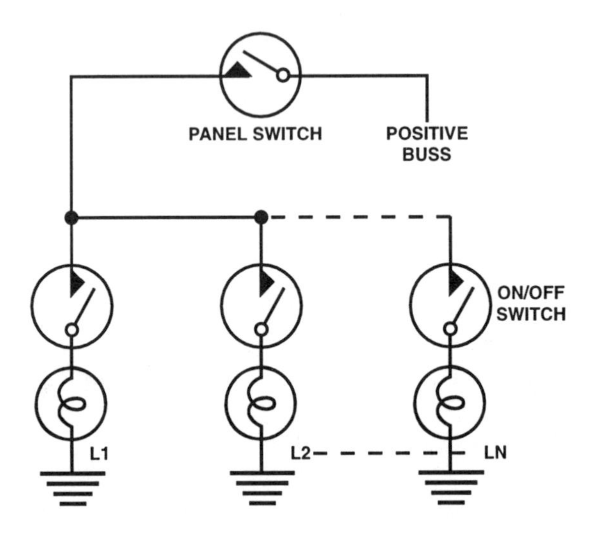

Figure 14.5: Cabin Lights

With two switches in series with a solenoid, which is in series with a motor, there is plenty of room for failure. The foot switch is particularly vulnerable if pin hole leaks develop in the rubber diaphragm. If possible, leave access to the foot switch for periodic inspections.

14.6 Cabin Lights

Figure 14.5 shows a circuit for various cabin lights. As usual, a switch on the control panel provides power to the lighting buss. Individual switches on the lamps turn the lamp on or off. The panel breaker or fuse should be rated such that all lights can be on at once. Naturally, the wires for the lights should be sized large enough for all lights as well.

14.7 Running Lights

Wiring for running lights is shown in Figure 14.6. All lights on the circuit are controlled by a single switch on the control panel. If you don't have a masthead tri–color light, then the compass light can also be activated by switch for the running lights. When a tri–color light is used under sail, a separate compass light switch can be used so that the compass light

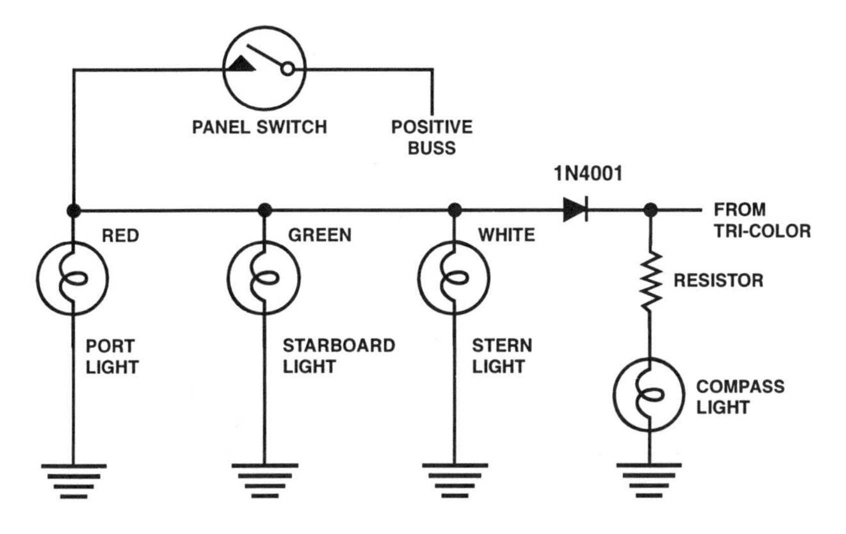

Figure 14.6: Running Lights

operates with either running lights or the tri–color. In Figure 14.6, we show a diode going to the compass light. With a diode coming from the tri–color circuit, a separate compass light switch is avoided.

The compass light has a resistor in series. Very little light is required to read a compass at night. As delivered by the manufacturer, our compass light could have been used as a stern light. We wired a 200 Ohm potentiometer in series with the light, and on our first night passage, adjusted the potentiometer for just enough light to see the compass clearly. Later we removed the potentiometer, measured its value, and permanently installed a 68 Ohm resistor.

14.8 Tri–Color/Anchor Light

A tri–color/anchor light is shown in Figure 14.7. Power is supplied through a switch on the control panel. The light assembly only requires two wires, and yet separate control is provided for either the anchor light, or the tri–color light. Which light illuminates depends on the position of a separate toggle switch. The toggle switch reverses the polarity on the two wires to the light. The two diodes in the light assembly are oriented so that one conducts for one polarity, while the other conducts for the

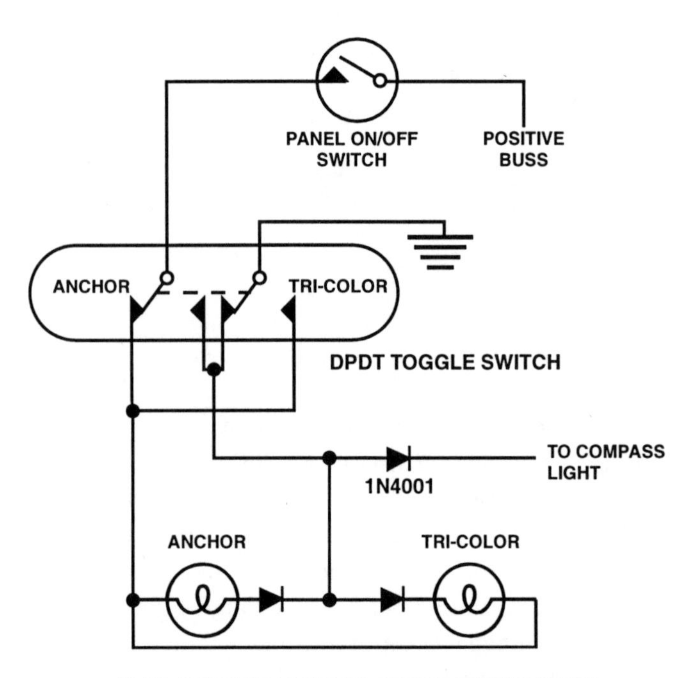

Figure 14.7: Tri-Color Light

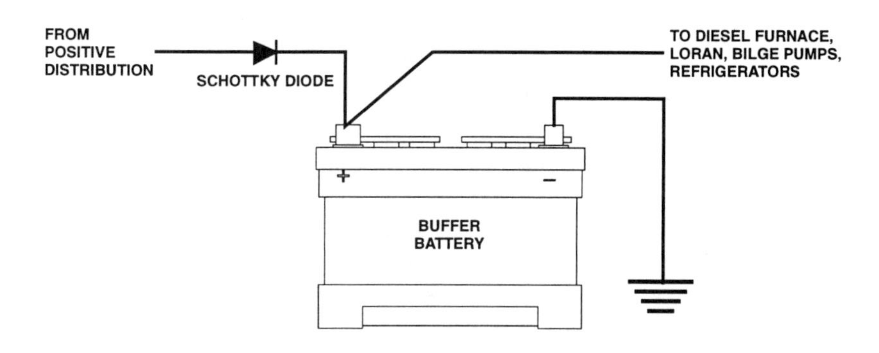

Figure 14.8: Buffer Battery

other polarity.

Some voltage is lost in the diodes, thus the lights will not be quite as brilliant as they would be without the diodes. Without diodes, three wires would be required. Besides the extra cost, more weight aloft results with three wires. If you decide to use a two wire assembly, you might consider replacing the diodes supplied by the manufacturer with low voltage Schottky units.

Note also the diode going to the compass light. Whenever the tri–color is on, the compass light is also powered. The compass light is shown in Figure 14.6.

14.9 Buffer Battery

Diesel furnaces, and some refrigeration systems should be connected directly to a battery. If a diesel furnace suddenly loses power when running, it can *meltdown* since it relies on a fan or water pump to cool. Small DC refrigerators have electronic drive circuits for their motors. If they loose power when running, the drive circuits can be destroyed from motor transients. Direct connection to a battery reduces the likelihood that power will be suddenly lost.

With direct connection to one of your house batteries, the battery becomes dedicated, and will not get the rest that is needed to be able to translate voltage readings into capacity remaining.

Some electronic gear, such as Loran, is sensitive to low voltage transients when the engine is started.

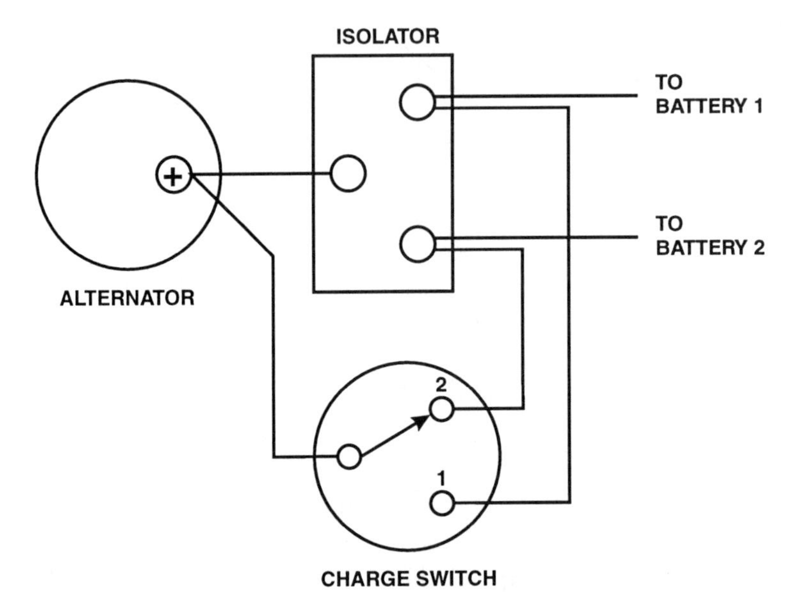

Figure 14.9: Isolator and Charge Switch

Figure 14.8 shows how a small *buffer* battery can be used to satisfy furnances, refrigerators, and electronics. The battery is charged from the positive distribution through the diode. The buffer battery needn't be large since it is always fed from the selected house battery. If the selector switch is turned off, the battery will discharge, however, low voltage circuits in both the furnace and the refrigerator will activate to protect the devices and the battery.

The diode feeding the buffer battery should be rated about 100 Amps. The reason it needs to be this big is to carry charging currents that will flow if the buffer battery ever gets deeply discharged. If you use a large capacity buffer battery, you will need a diode that is rated for more current. The diode should be a Schottky type with low forward voltage drop. Heatsinking will be required on the diode. Refer to *Living on 12 Volts with Ample Power*, for data about choosing adequate heatsinks.

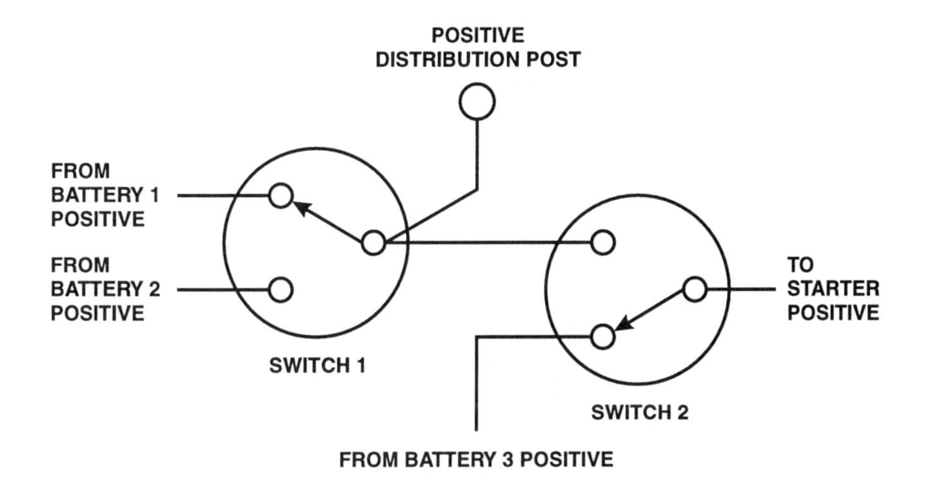

Figure 14.10: Three Batteries

14.10 Isolator and Charge Switch

Isolators are convenient since you don't need to know which battery needs the most charge, and you don't have to worry about putting the switch on both and forgetting that both batteries are connected. With isolators, you also don't have to worry about someone turning the selector switch off with the alternator running.

Sometimes it is convenient to direct the charge to one battery without interference from the other. Figure 14.9 shows how a selector switch can be wired with an isolator. With the selector switch in the off position, the batteries are charged through the isolator. A leg of the isolator can be shorted by the switch. This will direct the charge to the selected battery.

14.11 Three Batteries

Figure 14.10 shows how three batteries should be wired with two switches of the 1–2–Both variety. Switch 1 selects one or both of the house batteries as normal. Switch 2 controls power to the starter. The engine can be started from any combination of batteries. Table 14.1 shows the switch positions that select the various batteries for starting power. If Switch 2 is off, then it doesn't make any difference where Switch 1 is.

Switches On		Battery Selected		
Sw 1	Sw 2	B1	B2	B3
Off	1	No	No	No
Off	2	No	No	Yes
Off	Both	No	No	Yes
1	1	Yes	No	No
1	2	No	No	Yes
1	Both	Yes	No	Yes
2	1	No	Yes	No
2	2	No	No	Yes
2	Both	No	Yes	Yes
Both	1	Yes	Yes	No
Both	2	No	No	Yes
Both	Both	Yes	Yes	Yes

Table 14.1: Three Battery Selection

14.12 Ship/Shore Switch

Figure 14.11 shows wiring for a Ship/Shore AC selector switch. As a minimum, the switch must be a double pole, double throw toggle switch as shown. Often, a rotary switch is used with a third *Off* position. There is a wider selection of current ratings for rotary switches. Toggles switches rated above 15 Amps are difficult to find, whereas 50 Amp rotary switches are fairly common.

14.13 Ship/Shore/Inverter

Figure 14.12 expands on Figure 14.11, by adding circuits for an inverter. Ship or shore selection is accomplished as before, but the output from the first switch is fed to one input on a second switch. The other input of the second switch comes from the inverter.

It should be noted that the second switch need only be rated for the current that all of the inverter loads can consume, not what the generator

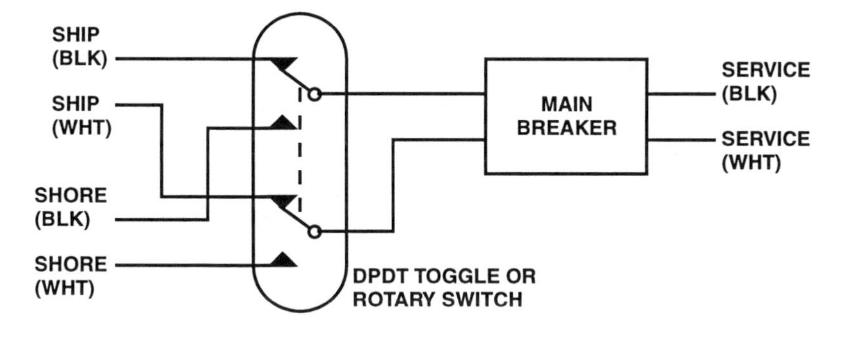

Figure 14.11: Ship/Shore Switch

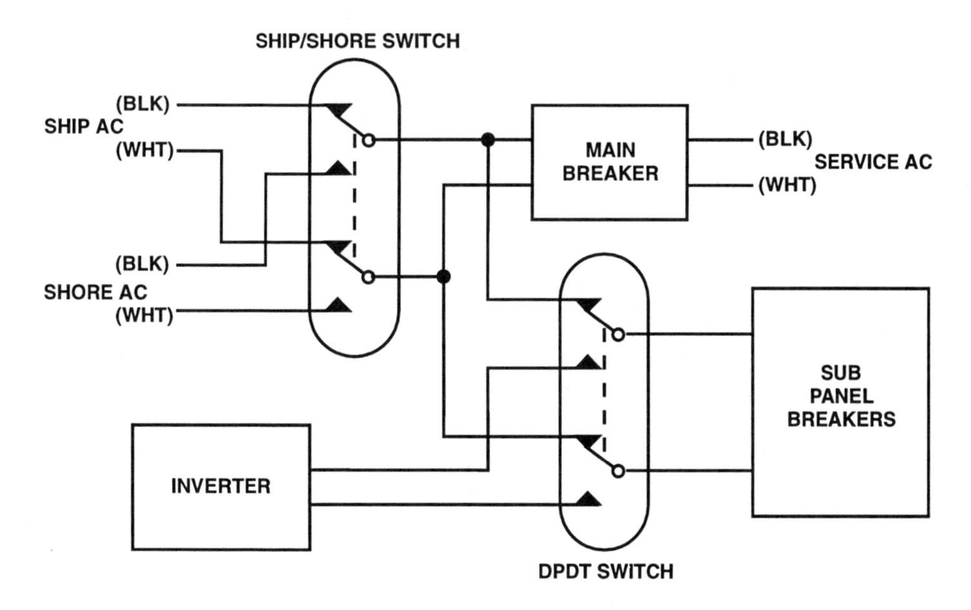

Figure 14.12: Ship/Shore/Inverter Switches

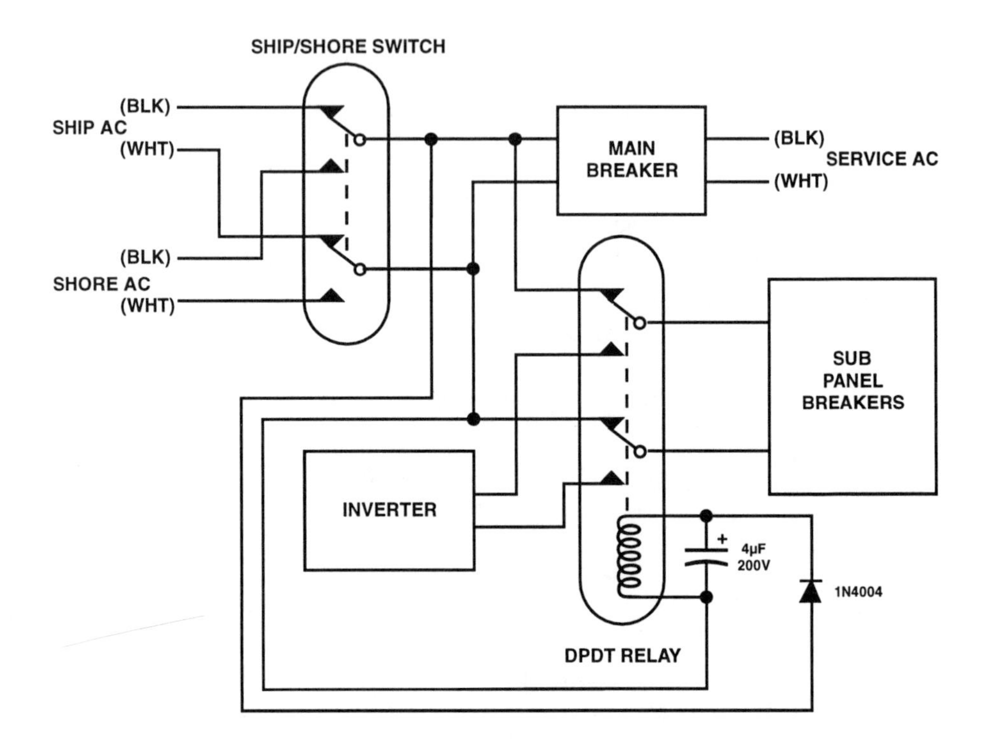

Figure 14.13: Inverter Transfer Relay

or shore can supply.

The second switch can be a relay. The relay is activated when either ship or shore power is available, automatically de–selecting the inverter. Figure 14.13 shows how a relay is used instead of a second switch. A 115 Volt DC relay is used, with a diode to rectify the AC, and a capacitor to store the charge. By using the diode and capacitor, the relay won't chatter whenever the generators starts or stops.

A capacitor of 2 micro–farads at 200 VDC is sufficient. The diode shown is rated at 1 Amp and 400 Volts.

14.14 Refrigeration Controls

Refrigeration control is shown in Figure 14.14. A thermostat senses temperature and closes a switch when the temperature is too high. If the timer switch and the panel switch are both on, then the compressor mo-

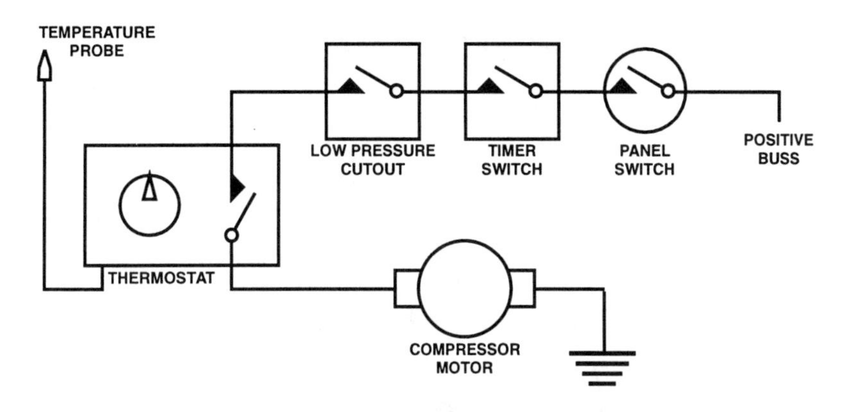

Figure 14.14: Refrigeration Controls

tor runs. The motor will stop running if the timer switch opens, or the low pressure cut–out switch opens. Low pressure is reached when the evaporator gets sufficiently cool. A high pressure switch in series with the other switches is sometimes included as an extra safety mechanism.

14.15 Instrumentation

Figure 14.15 shows how easy the BatMon from Ample Technology[1] can be wired. It consists of two assemblies, the shunt assembly and the display assembly. Interconnection is done with a telephone type cable.

The shunt assembly is connected in series with the negative wire from the battery. A small positive wire from the shunt assembly goes to battery positive to pick up power and sense voltage. The BatMon displays Volts, Amps and Amp–hours consumed.

14.16 AC Wiring Schematics

There are three common AC capacities supplied at marinas and RV parks, 15 Amp, 30 Amp, and 50 Amp. Sometimes 20 Amp service is found. You should really only plan on drawing about 80% of rated capacity from any AC service . . . 12, 24, and 40 Amps respectively. The sockets used for AC

[1] Ample Technology, Seattle, WA

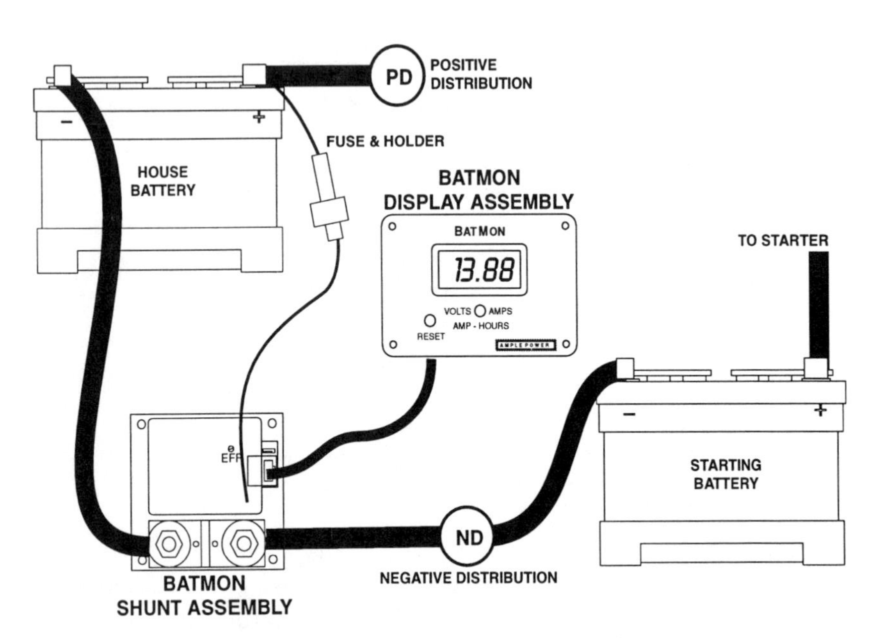

Figure 14.15: Instrumentation Wiring

distribution will overheat if loaded to capacity. As the socket gets older the blades won't grip the plug as tightly and the resulting poor connection contributes to heat buildup. We're not suggesting this as a remedy for loose fitting sockets . . . they should be replaced, but some people take a couple pair of pliars and *dish* the plug a little to make a better contact. Of course that spreads the socket blades even more and contributes to even less contact pressure for normal flat blades.

There are three wires used for 15 and 30 Amp service, a hot wire, a neutral wire and a safety wire. At some point on land, the neutral and safety wires are connected together and to earth itself. **NEVER** connect the two wires together yourself at any point. You'll defeat the safety function and may cause injury or death.

A house is normally wired with 15 Amp outlets. These same outlets are supplied at older marinas and RV parks.

Figure 14.16 shows a 15 Amp socket as seen at the junction box. Shown also are the 3 wires mentioned earlier. The neutral wire is shown as a line, while the hot line is shown as a sinewave that alternates positive and negative with respect the the neutral wire. Note that the sinewave

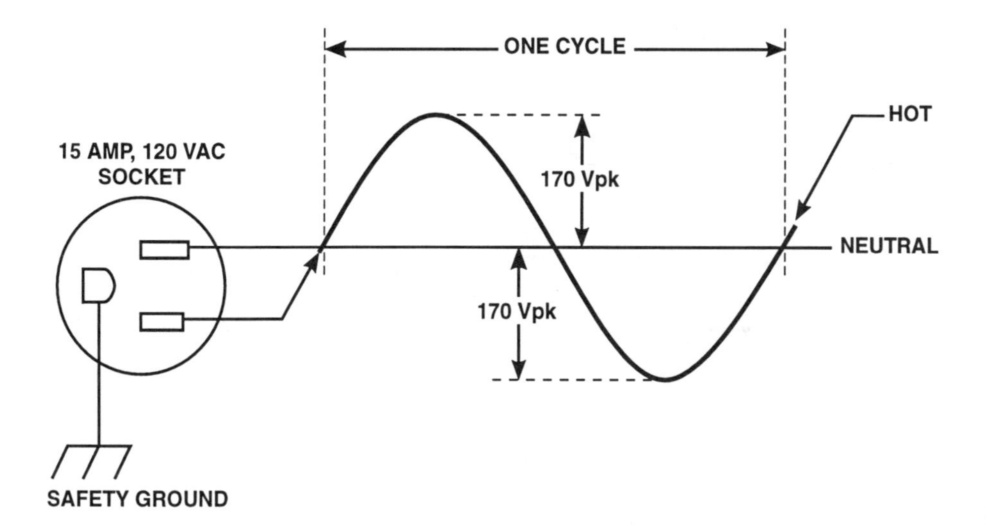

Figure 14.16: 15 Amp AC Socket

has 170 Volts peak. You may recall that AC is rated in RMS values which are 0.707 times the peak value of a sinewave . . . in this case 120 VAC.

Thirty Amp service is also provided with the same three wires as 15 Amp service. The socket is mechanically different to prevent interchange with other service capacities. It also has bigger contacts to carry the additional current.

Figure 14.17 shows the configuration of 30 Amp service. Again, the socket is shown as it would appear externally at the dock or park hookup.

Fifty Amp service is normally provided as 240 VAC between two hot legs. A neutral wire is provided such that each hot wire to neutral is 120 VAC. This is shown in Figure 14.18. Note carefully that the sinewaves for the two hot wires are out of phase. If they were in phase, there would be no net voltage between them. Sometimes, 50 Amp service is wired this way because an unqualified electrician did the wiring. There would still be 120 VAC from each hot leg to neutral, however, so if all you used is 120 VAC service you might not notice.

There is a subtle problem with the in–phase mis–wire. Each hot leg is rated for 25 Amps. If you were using 25 Amps from each hot leg, for 50 Amps total, how much current is flowing in the neutral leg. If the circuit is wired properly with out–of–phase hot legs, then the neutral leg

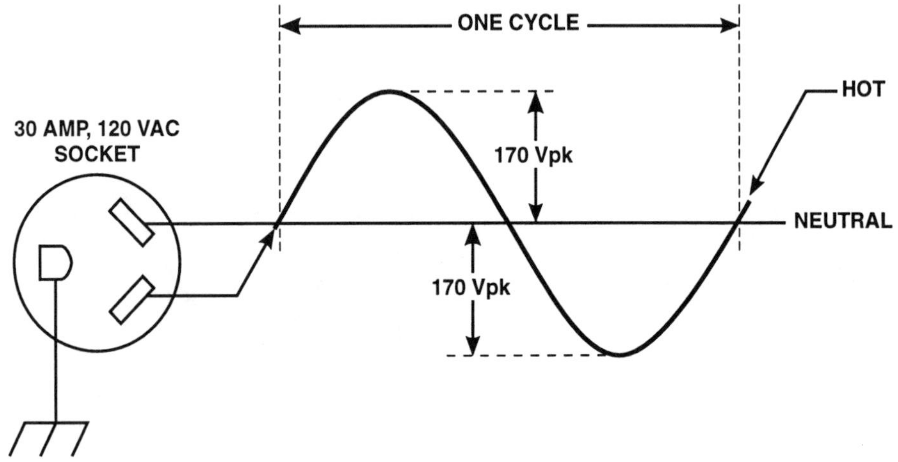

Figure 14.17: 30 Amp AC Socket

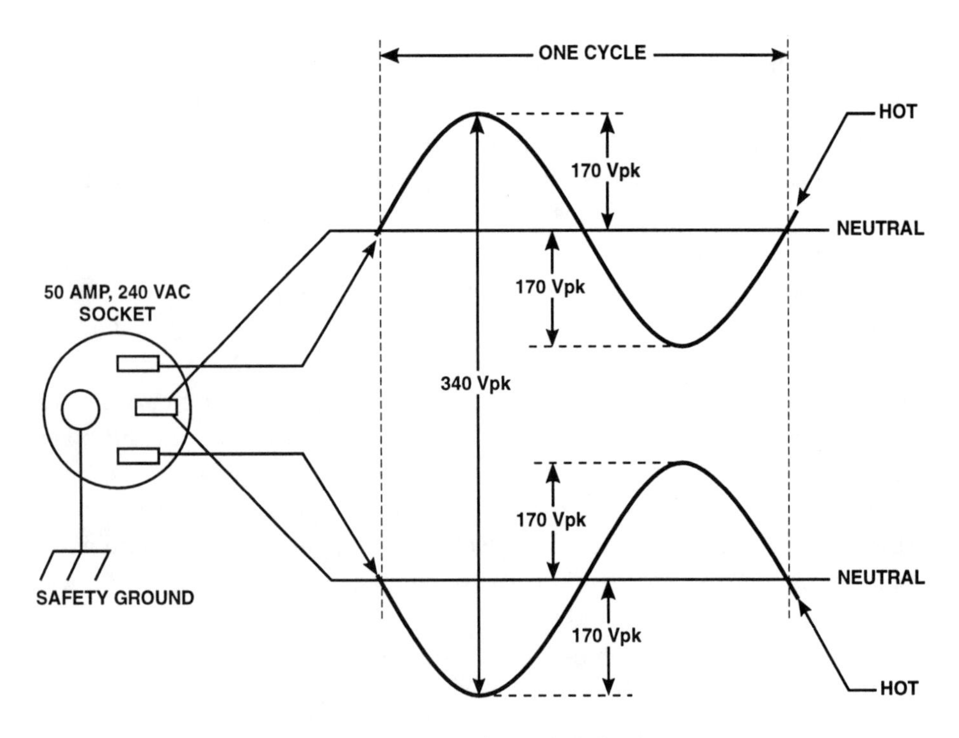

Figure 14.18: 50 Amp AC Socket

has NO current in it! If the circuit is mis–wired with in–phase hot wires, then the neutral wire has to carry 50 Amps! The neutral wire is probably only rated to carry 25 Amps since this is the most it would have to carry in a correct circuit. That happens if only one hot leg is conducting at 25 Amps.

We've seen power cords that have burned up the neutral conductor from such mis–wired 50 Amp service. Power cords are available with a neutral conductor rated for 50 Amps, but is your internal wiring?

In too many instances, the wires of 15, 30 or 50 Amp service get crossed. Hot and neutral swaps are the most usual. This is called reversed wired and represents a serious shock hazard. There should never be much of a voltage between neutral and safety ground. We've seen about 3 VAC at a marina, which we thought was quite high.

Don't even think about plugging into an unknown socket without first testing it for correct wiring. An unknown socket is one which you haven't personally tested. Once we left with our boat on vacation and when we returned, discovered that the socket at our slip had been rewired while we were gone . . . hot and neutral had been reversed. The marina office denied having rewired the slip . . . perhaps someone decided to play a potentially deadly joke.

Note that if you wire your boat or RV with 50 Amp service there will be two internal hot circuits representing the two hot wires. Typically you want to balance the loads between them. For instance, if you have two air conditioners, wire one to each hot leg.

Inverters generally provide only single phase AC. How do you use an inverter with the two hot legs? You have to short the two legs whenever the inverter powers them. This can be avoided by concentrating the loads to be inverter powered on just one of the hot legs. Wired in this manner the inverter only has to supply one of the hot legs, not both.

Chapter 15

Wiring Practices

15.1 General Information

Wiring practices include such subjects as choosing sites for electrical apparatus, planning the wire runs between apparatus, and making the actual hook–ups. Also considered are wire size requirements, wire identification, and anchoring wires to prevent chafe.

15.2 Wire Sizes

The most important wiring practice is to observe proper wire size. Failure to use adequate size can result in fire. Even if fire doesn't result, wires that are too small will cause marginal performance of electrical equipment.

Table 15.1 shows the wire size required for a 3% voltage drop *in 12 Volt* circuits. To use the table, first calculate the total length of the wire from the source to the device and back again. Next, determine the amount of current in the wire. The wire gauge is found at the intersection of Amps and Feet. In most load circuits, a 3% drop is quite acceptable. In charging circuits it often pays to have less of a drop.

Table 15.2 gives the cross sectional area of wire gauges. Voltage drop at the load can be calculated from Amps and circular mils using the following equation:

$$E = (10.75)(L)(I) / C$$

	Distance – Feet						
	10	15	20	25	30	40	50
Amps	Wire Gauge						
5	18	16	14	12	12	10	10
10	14	12	10	10	10	8	6
15	12	10	10	8	8	6	6
20	10	10	8	6	6	6	4
25	10	8	6	6	6	4	4
30	10	8	6	6	4	4	2
40	8	6	6	4	4	2	2
50	6	6	4	4	2	2	1
60	6	4	4	2	2	1	0
70	6	4	2	2	1	0	2/0
80	6	4	2	2	1	0	3/0
90	4	2	2	1	0	2/0	3/0
100	4	2	2	1	0	2/0	3/0
120	4	2	1	0	2/0	3/0	4/0
140	2	2	0	2/0	2/0	4/0	4/0
160	2	1	0	2/0	3/0	4/0	4/0+4
180	2	1	2/0	3/0	3/0	4/0+10	4/0+2
200	2	0	2/0	3/0	4/0	4/0+4	4/0+0

Table 15.1: Wire Size versus Distance and Current

Wire Gauge	Circular Mils
18	1600
16	2400
14	3800
12	6000
10	10,000
8	16,000
6	26,000
4	40,000
2	64,000
1	80,000
0	100,000
00	130,000
000	164,000
0000	208,000

Table 15.2: Wire Gauge Area

AWG/MM Size Conversion				
AWG	MM	–	AWG	MM
26	.12826		11	4.156
25	.162		10	6.271
24	..205		9	6.626
23	.255		8	8.350
22	.322		7	10.544
21	.411		6	13.292
20	.516		5	16.755
19	.653		4	21.137
18	.823		3	26.653
17	1.039		2	33.606
16	1.308		1	42.384
15	1.652		0	53.454
14	2.088		00	67.399
13	2.629		000	84.004
12	3.302		0000	104.091

Table 15.3: Conversion from AWG to Millimeter Sizes

In this equation, E is the voltage drop, L is the length of wire to and from the device, I is the current in Amps, and C is the circular mils from Table 15.2. You can rearrange the equation as shown below to give circular mils, given an acceptable voltage drop.

$$C = (10.75)(L)(I) / E$$

Any time you calculate circular mils, always use the wire gauge that is bigger than the minimum calculated.

For yet another way to determine wire size, refer to the book, *Living on 12 Volts with Ample Power.*

Many yachts go outside the U.S. and find that wire is specified in millimeter sizes. Table 15.3 is useful to determine appropriate sizes should you need to purchase wire.

15.3 Power Busses

Running individual wires to every device from a central control panel can significantly increase total wire length in a yacht. In some cases, it makes sense to run a few *power busses* that serve an assortment of equipment. Cabin lights that have on/off switches at the fixture can be run on a single buss, or a port buss and a starboard buss. Water pumps that work from a pressure switch can be connected to a buss.

When connecting equipment to a buss, it will help later to place quick disconnect terminals in–line at the device. The device can then be easily disconnected from the buss if fault isolation tests are necessary.

15.4 Choosing Sites

Choosing a location or *site* for electrical equipment can be complicated by a bewildering assortment of possibilities. In many cases, the location of an electrical apparatus is fixed by its function. Except in unusual circumstances, you probably won't mount a radar in the engine room, or a bilge blower in the owner's cabin. Other equipment may allow some choice as to its site. Such items as shunts, isolators, Charge/Access diodes and regulators may be mounted anywhere in the boat and still provide their intended function. Is there a *best place* for such equipment?

When choosing a site, it is important to follow the recommendations of the equipment manufacturer regarding temperature, moisture, air circulation, proximity to the battery, etc. Sometimes explicit restrictions exist which prevent mounting a unit in an otherwise perfect spot. The first step then, is to make a list of places where the equipment cannot be mounted. Many boatbuilders ignore this step ... all too often, the electrical panel is mounted below or adjacent to the companionway ladder where water splashes are guaranteed.

After ruling out certain places where equipment cannot be mounted, make a similar list of places where it can be mounted. If there are favorable or unfavorable attibutes of a given site, make a note of them. As you proceed, you will generate a matrix of possible sites for the various equipment in the wiring project.

Now you are ready for the most interesting part, choosing from the possible sites, those locations that minimize wiring runs. At the onset, be aware that wires can seldom be run from point to point, that is, floors and bulkheads present natural obstacles that must be circumvented. What often appears to be a simple straight run turns into an impossible maze.

Even if all wires could be run point to point, choosing sites to minimize wire lengths is a non–trivial problem. In fact, the problem of minimum interconnection distances is unsolvable by any presently known deterministic mathematical models, so don't expect that you will necessarily find a minimum solution.

In computer research circles, the minimum interconnection problem is often called the traveling salesman problem ...for a given group of cities, plan a minimum route that covers all the cities. Of a more practical nature, consider the economics of routing telecommunication wires that covers all points with minimum wire. For further reading of this interesting problem, refer to the January 1989 issue of *Scientific American*[1].

15.4.1 High Current Equipment

Despite the fact that experts have yet to find a method to plan minimum wire runs, finding a best fit is still mighty important considering the cost of wire. Since large wire is most expensive, not to mention heavy, minimizing long runs of large wire is the primary goal. Equipment connected with large wires include batteries, alternators, engine starters, inverters, anchor windlasses, and some battery chargers. Locating these items as close as possible will shorten the interconnections. Figure 15.1 shows the high current interconnections that must be made in a performance system. Efficiently connecting the equipment shown is a challenge that directly affects your wallet.

If you are planning a new boat layout, it makes sense to place the batteries close to the engine so that the alternator and starter wires are as short as possible. Mounting batteries in the engine room is often done, but battery life suffers if the engine room operates too hot. The best solution is to place batteries adjacent to the engine room, but out of the direct heat.

When a multi-battery isolator is used with the alternator, the isolator should be placed close to the batteries. Since one wire runs from the alternator to the isolator, while each battery must be connected to the

[1]Scientific American, Inc., New York, NY.

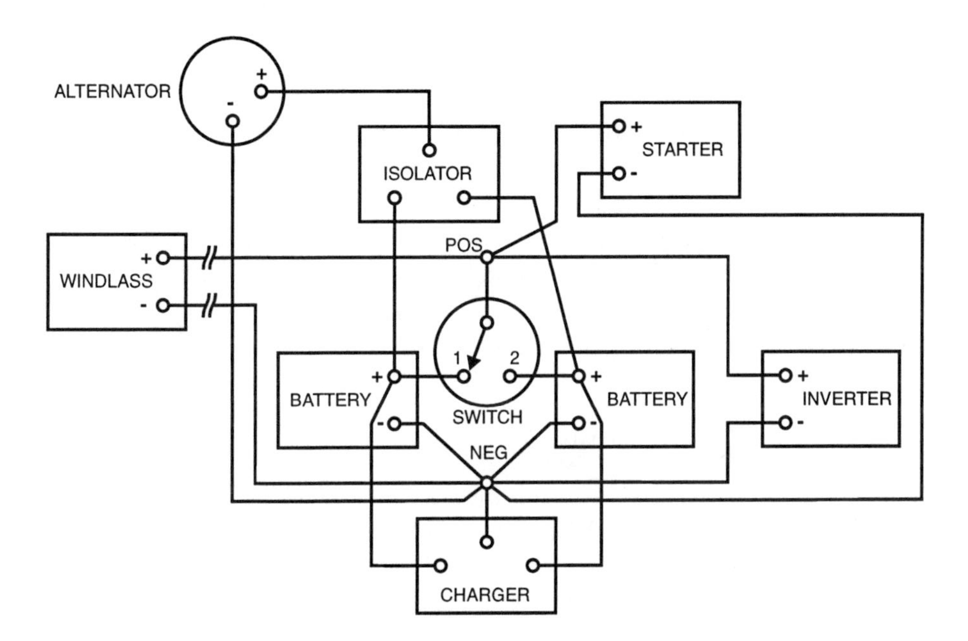

Figure 15.1: High Current Connections

isolator, wire length is minimized. On the other hand, isolators also generate heat which may shorten battery life. Of course if the batteries are not located together, then the isolator should be placed midway.

The battery selector switch is usually wired with large wire, especially if the engine starter is wired to the common of the switch. Again, since two wires go to the switch, with one output wire, it is best to locate the selector switch as close to the batteries as possible. A conflicting requirement is the fact that the switch needs to be easily accessible.

An inverter is a large electrical load. As such, it makes sense to place the inverter as close to the battery selector switch as possible. Keep in mind, however, that inverters generate heat, and may require ventilation if run for extended periods of time. Usually an inverter running from batteries doesn't operate long enough to generate much heat, but if you are using an inverter to power refrigeration during the time that the alternator runs, plan for plenty of cooling.

While it may make sense to mount an inverter close to the batteries, be advised that liquid electrolyte batteries generate hydrogen and oxygen during heavy charging. Not only is the mixture explosive, but oxygen also encourages corrosion ... electronic equipment should not be mounted in close quarters with batteries unless the batteries are of the sealed type.

The anchor windlass location is fixed by its function. Since two large wires must service the windlass, perhaps it makes sense to connect the inverter on the same wire run. You shouldn't run the inverter and windlass at the same time, but that is unlikely in any case. Sharing large wires is not limited to the inverter and windlass. Can a single positive wire service the engine starter, the windlass, and the inverter?

15.4.2 Electronic Equipment

Choosing sites for electronic equipment is not usually done to minimize wire length. The first criteria is to place the equipment where it is readily available to the navigator. In doing so, however, sensitivity to noise pick–up should be considered. For instance, a Loran receiver should not be located side by side with an inverter which is certain to generate noise.

When wiring any electronic equipment, it is good practice to *twist* the power leads. By twisting the positive and negative leads about one another, any noise pick–up tends to cancel. That is, the noise is induced equally in both leads, but because current is flowing in opposite directions in the positive and negative leads, the induced noise is out of phase and therefore cancels.

A small battery with a Charge/Access diode is often used to protect electronics from low voltages whenever the engine is started. The battery should be located as close as possible to the electronic gear.

15.4.3 Pumps

Choosing sites for pumps is not only dependent on the wires that connect them, but obviously on the hoses. Since hose is generally more expensive than wire, placing the pumps for minimum hose length is advised. Some pumps are *on demand*, that is, they normally operate via a float switch or a pressure switch. When pumps operate on demand, a separate on/off switch at the electrical panel is not always needed. In this case, the pump need not be wired with individual wires, but can be wired from a power buss.

15.4.4 Placing AC Equipment

We have an aversion to AC wiring aboard boats, and thus tend to minimize our usage of it. Nevertheless, AC appliances are useful, especially when connected to a dock. Choosing sites for AC appliances is most likely dictated by their function. A microwave oven is not usually found in a stateroom.

The key to successful use of AC aboard is to keep the AC wire runs short, high out of the bilges, and clearly identified. While a microwave logically goes in the galley, would it reduce AC wiring if it were located elsewhere?

For infrequently used appliances, it is best to avoid permanent wiring. When the appliance is needed, plug it into an outlet. Keeping one or more extension cords handy can simplify infrequent usage.

We shudder anytime we come across an AC outlet in a head. AC and water just don't mix. If you are using AC in the head, use a ground fault isolator.

15.5 Planning Wire Runs

15.5.1 Planning a Route

Most designers and builders appear to regard wiring as an afterthought. Consequently, running wire through a boat is usually the most time consuming wiring task. With floors to lift, bulkheads to avoid or bore through, and interior hull liners to circumvent, it can sometimes take several days or more just to run two wires. Since it doesn't take much longer to pull a bundle of wires, rather than just two, it pays to plan the wire runs for minimum effort.

If you are building your own boat and are now ready to wire it, you fare much better than the hapless buyer of a boat designed by a *wireless* architect. As a builder, you know every surface in the boat, and whether it is conducive to a wire bundle.

As an innocent buyer who has elected to upgrade the electrical system, your first step will be to learn what possibilities exist. The first thing to look for is wires, as they presently lay. If the builder was conscientious, the wire runs will be obvious. Furthermore, the builder will have left in place a pilot line that is used for pulling more wires through areas not readily accessible. **It is good practice to always pull a new pilot whenever adding a wire to the system.**

Many owners will find that builders wire the boat in devious ways. When a boat is wired prior to finish of construction, and the builder leaves no pilot, you will have to cut an existing wire and use it as a pilot. You will have to replace the cut wire, of course.

When looking for wire routes through territory not presently wired, only ingenuity appeases frustration. How does the liner come free? If you don't find screws, look for Velcro[2] or other hidden fasteners. You may have to penetrate a bulkhead with a hole saw to get from A to B. Before laying any wire, plan the route, and make all accesses. Nothing is more frustrating than pulling wires out because you hit a blind alley.

15.5.2 Drafting a Wire Routing Plan

By now, sites have been chosen for the equipment, and wire routes are open. It is time to draw a wire routing plan. While any paper will suffice, it makes sense to draft the plan on vellum which can then be reproduced. Check with an engineering/architectural supply for reproducible paper such as Clearprint[3] 1000HP. Drafting paper comes with or without dimensional grids. Grids can be either fractional such as 8 to the inch, or decimal such as 10 to the inch.

If you haven't used reproducible paper before be advised that the two sides have different textures. On Clearprint paper, draw on the side that *does not* have the grid marks. Drawing on the grid side will leave smudges whenever you erase.

The first drafting step is to make a flat outline of the boat looking at it from topside in. The outline doesn't need to be pretty, just communicative of the interior layout. At each equipment site, draw a rectangle and label it with a mnemonic. For example, the inverter might simply be labelled **INV**.

On the drawing, connect all sites via the routes available with a heavy line that represents a wiring bundle. At each site, indicate the number of wires that connect.

[2] Registered trademark of Velcro Inc.

[3] Clearprint Paper Company, Emeryville, California.

After counting wires that terminate at each site, determine the destinations and make note of them on the drawing. As an integrity check, add up the wires to each *destination*. The total of destination wires had better be equal to the total wires which terminate at the site.

The next step in the planning process is to determine the gauge of each wire. Wire gauges should be placed on the diagram.

To determine wire size, wire length must generally be determined first. Length can be estimated with a few rough measurements, or it can be measured quite accurately by pulling strings through the routes and measuring the string length. In either case, allow some slack ... wire runs get longer as wire is placed. Record the wire length on the wiring plan.

Where possible, the wiring plan should also include a means of wire identification. Wires can be identified with tags after they have been pulled, but wire tags are generally not possible during the actual routing. Where you plan to run several wires at once, it is good practice to color code the wires.

While color coding wire is a good means of identification, it has a caveat ... some wire colors may not be readily available, and purchasing several short lengths of various colors may be more costly than a longer length of a single color.

Site labels with a destination, wire gauge, length and identification are strictly a convention. Use anything that makes sense, and can be communicated to the person that will actually route the wires. If that person is yourself, it is easy to cheat a little on the drawing. Avoid the urge ... when you re–visit the plan later to add more gear, a clear and accurate plan will pay off.

If you have drafted your wire routing plan on vellum, make several blueline copies before actual wire routing. One blueline should be labelled as the official work copy ... any changes made during installation should be clearly marked so that changes to the original can be made later.

Most boats do not have good, *as wired* diagrams. The owners pay a premium anytime the wiring needs to be troubleshot, or additions are made. By making a clear and accurate wiring plan, you will have increased the value of your yacht.

15.5.3 Material Acquisition

With the wiring plan ready, you may purchase the wire and terminals that are required. It helps to prepare a materials list. Whether you purchase wire from a marine store, or from a local electronics supply, a formal material list will be necessary. You can even use the list to price the materials from several possible sources.

Besides wire and terminals, you will want some way of tying wires into bundles and anchoring the bundles. Wire markers will also come in handy. It isn't always possible to plan the number of anchors or ties that are required, but one per foot is a good place to start. These items are usually cheap enough to buy a few extras if it will prevent a trip to the supplier later during installation.

15.5.4 Running the Wires

There are two basic ways to tackle the task of pulling wires between sites ...everything at once, or site by site. To pull all the wires at once may not be the best idea if you are new at wiring, but in some circumstances can be quicker than pulling one site/destination at a time.

Pulling all wires at once can not usually be accomplished if you must make extensive use of a pilot. There are limitations to how many simultaneous wires can be routed through a difficult passage.

To run all wires at once, start at one distant point, and work to the point furthest away. As you route past equipment sites, you will have to drop wires that terminate at the site, and pick up other wires that lead to other sites. As you proceed, place tie anchors, and temporarily tie the wire bundle in place. If you think all wires are in place, go ahead and make the tie permanent ...frankly, we usually leave permanent wire tying until after all equipment is hooked up and operating.

For any complex wiring task, it is usually best to pull wires from one site to another according to their function. The first wires to run are the large power wires. By pulling them first, you avoid snarling the smaller wires that run adjacent, and routing large wires is easier when there are fewer obstacles.

As you run the first wires, place a few tie anchors to loosely hold the wires in place. We call these the *routing anchors*, and usually place them where the wires must make sharp bends. Routing anchors are also advised where long straight runs are made. Place the routing anchors close enough to avoid large sags in the wires. It is often easier to place anchors before the wiring bundle gets too large. Only a few of the anchors are used with temporary ties as the routing continues, but the rest are ready to be used when wire routing is complete.

Routing a few wires at a time has the advantage that wires can be identified after they are in place. Wire tags can be placed on individual wires, or on a whole bundle. Good identification will pay dividends at the time of termination. While you often have to *buzz for continuity*, well identified wires can save alot of needless searching for the right wire. To identify wires, either use a permanent wire tag, or simply wrap the wire in masking tape and write on the tape with an ink pen. You can also use a permanent marking pen with a fine point to write directly on the wire.

Some people are successful in terminating wires as they are run, thus avoiding the identification problem. Concurrent termination often leads to a sloppy looking wire bundle, however.

As noted earlier, permanent wire ties can be placed whenever you are confident that all wires are in place. For neatest appearance, wires should be bundled and tied prior to hook–up. By proceeding in this fashion, excess wire length can be worked toward the point of termination and trimmed to length at that time.

You can avoid permanent ties before hook–up and test, and still achieve a reasonable appearance by cinching only the ties at routing anchors, and placing a permanent tie at the point of termination. There are two traps here. First, permanent wire ties may actually increase the distance between sites. This is particularly true if you have not placed routing anchors at all the corners, and cinched them up prior to termination. Since stretching wires to make a neat bundle is impossible, be sure that there is plenty of wires left at the termination point.

This brings up the second trap ...excess wire length. Excess wire can be unsightly. While some excess wire is a good idea, it should not be visible at the point of termination. At the time of termination, fasten the wire bundle close to the actual terminals. Leave slack between the terminals and the wires leading away from the site. After termination, any excess wire is *squeezed* away from the termination where it can generally be hidden in the bundle.

15.6 Connection to Pilot

It can be sheer frustration to lose a wire bundle midway through a difficult passage. In fact, you may not be able to replace the Pilot easily (more on this subject later). There are several methods of connecting a pilot to a bundle. Only two of them have real merit.

The way not to make the pilot to bundle connection is with tape. Sure, tape will work in many cases, but the tape method follows Murphy's law exactly ...it always fails at the most inopportune time.

For most circumstances the pilot can be a heavy duty string such as the waxed lacing cord discussed elsewhere in the book. The way to fasten it to the bundle is with a *rolling hitch* shown in Figure 15.2. A rolling hitch is quite secure ... we have even used the knot to fashion rope steps up a stainless steel shroud. The more rolls before making the cross over, the more friction the hitch has.

After making the rolling hitch to the wire bundle, place tape over the hitch, and then tape toward the bundle end. The idea is have the cord exit the bundle about at its center, with tape forming a pointed end that won't hang up on obstacles inside the passage.

When you must traverse a difficult passage, string or lacing cord may not be strong enough. In this case, a small wire pilot such as #14 gauge is called for. While a rolling hitch may be appropriate with a wire pilot, it is best to strip the insulation from the pilot and the wires in the bundle and then solder the wire to the bundle. The major concern with soldering the pilot to the bundle is in the strength of the solder joint. For best results, first wrap the pilot around one of the wires in the bundle and solder it to the single wire. Now wrap the other wires in the bundle around the junction and solder them. As usual, wrap tape around the bundle to form a fair leading edge at the end of the bundle.

Note: always include a new pilot in the bundle.

15.7 Soldering

Soldering is relatively easy, but does take practice before you can achieve good junctions. For best soldering work, the items being soldered must be mechanically bound together before application of solder. Relying on a solder joint that was done without first making a good mechanical

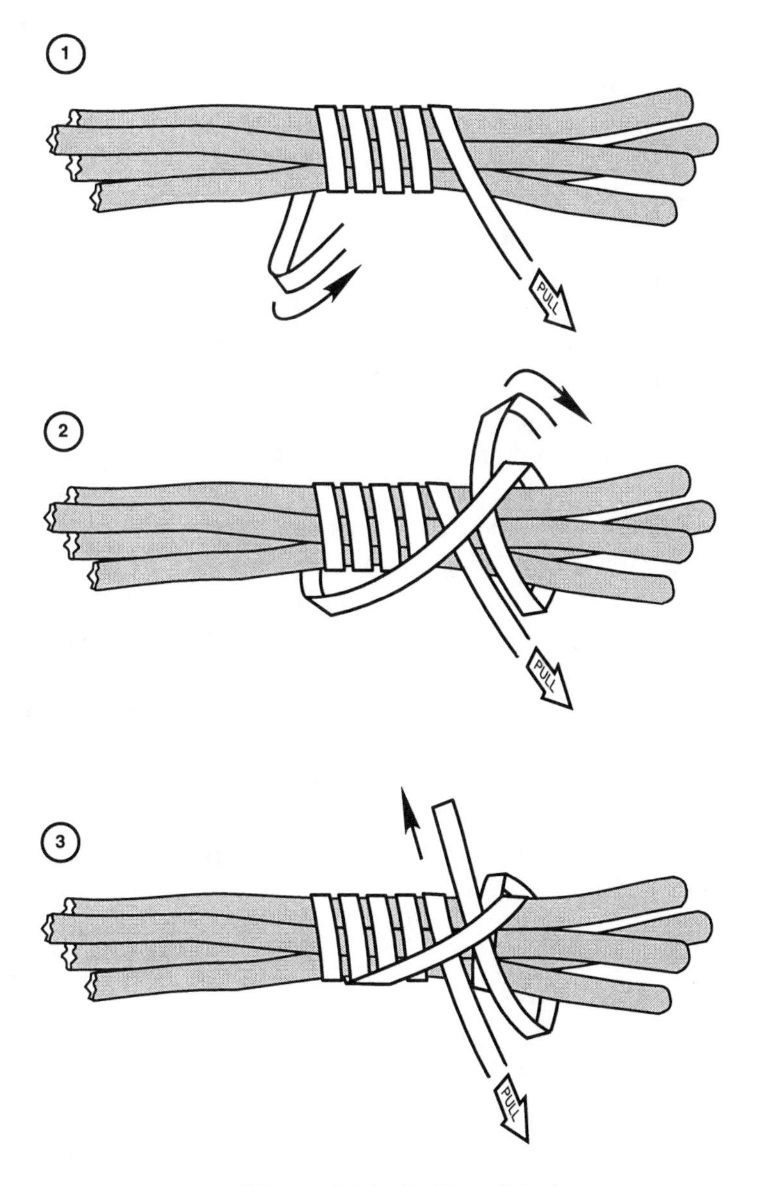

Figure 15.2: Rolling Hitch

connection is risky, since solder by itself is easily stretched apart.

Solder needs to cool to a solid without undergoing any motion as it cools. Motion upsets the cooling process and results in joints that are brittle and easily fractured.

Solder must flow readily across the joint. If you make joints that have clumps of solder, rather than a single homogeneous flow, then the junctions will separate in time.

When soldering, first clean the tip by dragging it through a wet sponge. The tip should be bright and shiny. Now make contact between the soldering iron and the work, allowing it to heat. Put a small amount of solder right at the junction of the work pieces. Do not apply solder to the soldering tool, instead, apply solder to the work itself. The solder should melt, and aid in the transfer of heat from the iron. Avoid the tendency to apply too much solder before a smooth flow begins.

Splicing two wires with solder is quite easy. First, cut a few single strands of wire about six inches long from some scrap wire. Strip insulation from the wires to be spliced about an inch back. Push a couple pieces of heatshrink tubing on the wires to be used later as insulation.

Gently push the loose wire ends together, getting as much cross penetration of wires as possible. Now lash the splice together with the previously prepared strands. You need to wrap the splice well enough so that it will stay together. If you can, use a clove hitch with the strands, tightly joining the wire splice. Now you can add solder to complete the splice.

After the splice cools, nip any wire ends that are exposed, and slide the heatshrink tubing into place.

15.8 Terminations

Many terminations are made to terminal blocks, either with the plain wire end, or via terminals. Other terminations may be made via connectors, or in the case of batteries, with large wire terminals.

Terminal blocks that connect directly to wire without a terminal are gaining in popularity. These types of terminal blocks provide an excellent long term connection without the possibility of a poor crimp or solder terminal. When using a terminal block that interfaces directly with the wire, *do not* tin the wire with solder before connection. Direct wire insertion terminal blocks are designed to squeeze down on bare stranded wire. As the force is applied to the wire, it tends to spread and thus widen its contact with the screw post. A soldered wire will be hard, and the actual surface area of contact between the wire and the terminal lug will be less than the bare wire connection. The reliability of a connection is directly proportional to the surface area of contact, and how *gas tight* the connection is. With sufficient force against bare stranded wire, many parallel gas tight connections are made. With increased surface contact, less heating due to connection resistance will occur. This in turn will maintain the connection gas tight, preventing oxidation from occurring.

Terminal blocks that are designed for use with a forked or ring terminal are popular, and very effective. As noted earlier, the terminal blocks should have hardware that is not magnetic. Steel screws will eventually corrode and should be avoided. Terminal blocks designed for use with wire terminations should never be used with bare stranded wire, or, with solid wire or stranded wire that has been tinned so as to make a solid wire. Use only the proper size of terminals.

A terminal should never be just soldered to the wire. If you do plan on soldering, crimp first. After soldering, protect the joint with tape or heatshrink tubing. Heatshrink tubing with an internal silicone sealant offers the greatest protection.

Some people insist on crimp/soldered connections. While our experience indicates that crimp terminations are quite sufficient, a post crimp soldering step may be called for, especially for connections that are deep in the bilge and subject to high humidity that surrounds standing water. Be advised, however, that soldering stranded wire essentially makes it a solid wire. Solid wire is much more apt to break with repeated flex. If you do solder terminals, be sure to protect the joint for any motion. Sometimes you can stiffen a wire/terminal joint with an extra layer of heatshrink tubing. You may also want to add another wire anchor close to the joints with a secure tie. Note that the U.S.C.G. does not allow joints to be only soldered.

Wire and terminal should always be crimped using an appropriate crimp tool. Don't even think about using pliers since they will not repeatedly make a good crimp. Crimps should be able to withstand a pull of about 5 pounds. If you aren't using a calibrated crimp tool then every crimp connection should be subject to the pull test. Even a calibrated crimp tool may get out of adjustment. Periodically performing a pull test on crimps made with a calibrated tool is good practice.

Prior to buying battery cable crimpers, we would use a vise to flatten the terminals over the wire. Then the terminal and wire would be soldered with complete saturation of solder. At least two layers of heatshrink tubing was then applied. While these junctions were not the prettiest, and were time consuming to make, none of them have failed. For an amateur method, the vise crimp and solder is appropriate where only a few connections are to be made.

15.9 Twisting Wires

As noted, noise reduction can often be accomplished by twisting two or more wires together. While it is possible to twist wires by hand, it is a boring job with only mediocre results. For a professional appearance, use a variable speed drill motor. Chuck up the wires tightly in the drill motor and either tie them to a distant point, or have a helper hold the wires. Now, keeping wire taught, slowly run the drill motor until the wires are tightly wrapped.

Once wires are twisted, *do not* release them at either end without stretching them tight. This helps to bind the twist together, and prevent a massive snarl when tension is removed. Release tension on the wires slowly to avoid snarling.

15.10 Testing Wire Runs

Despite the certainty with which you identified wires prior to termination, it is good practice to test wire connections prior to activating any equipment.

Sometimes manufacturers provide voltage measurement tests which can be performed to assure proper hook–up. More often, however, you will have to make continuity tests of the individual wires. For this you will need a long wire, an ohmmeter, and a helper. With the helper connecting one end of the long wire to the termination under test, you will measure continuity between the long wire and the corresponding termination at your site. Each wire must be *buzzed* or checked for continuity.

15.11 Making Splices

Splices are sometimes a necessary evil. One way *not* to make a splace is with twist–on wire nuts commonly used in household wiring. They will eventually loosen from vibration.

Crimped butt connectors make a fast splice. They should always be sealed with heat shrink tubing, or silicone with a wrap of tape.

Soldered splices are permanent, if the wires are not subject to vibration. As mentioned, soldering stranded wires can lead to brittleness and the chance of a wire breaking. Always support soldered terminals to avoid continual flexing.

If several splices are required at the same place, use a terminal block.

15.12 Crimping Big Lugs

There is no substitute for calibrated crimping tools, but their cost is prohibitive if all you want to make is a few crimps. If you don't have a crimp tool, you can use a vice to make some mechanical connection. A vice flattens the terminal on the wire, so mechanical strength is not high. The terminal should be soldered after the crimp has been made.

To solder the terminal, secure it in a vice, or in vice grips, with the wire exiting the terminal from the top. Heat the terminal, not the wire, with a small soldering torch. The wire will wick up alot of solder, but the idea is to get the solder into the terminal, not up in the wire.

Chapter 16

Testing Electrical Systems

16.1 Introduction

Finding the source of failures in electrical systems is generally a process of discovering where power or ground wires have opened up. Many times, however, failures can be more subtle. In particular, intermittent problems cause much grief.

While we would like to be able to present an exhaustive table of diagnostic steps for every imaginable failure, a finite lifetime doesn't make that possible. We attempt instead to give a few ideas that will be useful in the event of failures.

There is no substitute for historical data. Before you have problems, take some time to snoop around in your electrical system. If you don't have good documentation, trace out the wires and make a drawing of the system. Measure voltages and currents with equipment idle, and again with it operating, recording the values. If components are warm, record the operating temperature as well. When trouble does strike, you'll be prepared.

16.2 Safety Considerations

High voltages kill! High DC voltages are not as lethal as AC, because a DC voltage tends to knock you away from the circuit. AC voltages weaken muscles, so you may not be able to release your grip on an AC wire, or shove yourself away from an AC circuit.

It takes very little current through the heart to kill ... a few thousandths of an Amp. With the heart in a direct line between your hands, you must avoid two hand contact with AC. Electricity can cause severe burns, so even if both hands aren't involved, you can lose a couple of fingers. Rings are particularly dangerous around voltages, even low voltages. You can easily melt a ring on your finger by shorting a battery. Consider yourself lucky if the molten metal doesn't splash in your eyes.

While DC voltages below 42 Volts are considered intrinsically safe, don't develop sloppy habits when working around low voltage circuits. Regard every battery with the same caution you would if it were a lethal power supply. Never touch both terminals at the same time.

With high voltage circuits, particularly AC, it is a good practice to make tests with one hand in your pocket. There is less chance of touching both sides of the circuit under test, if one hand is safely in a pocket or behind your back. Since you need to connect the meter to both sides, you will need a clip on one meter lead. That lead is clipped on with one hand, and then the same hand can be used to touch the other meter lead to the other terminal under test.

If possible, always attach the meter leads to the circuit under test with power off. Turn on power long enough to make the test, and then turn it off to remove the meter leads.

If you must test live AC or high voltage DC circuits, have a friend standing next to the circuit breaker to shut it down promptly.

16.3 Mechanical Failures

Many electrical failures are directly caused by mechanical failures, that is, the circuit opens up due to faulty connections. If you have an accurate wiring plan, all the terminal blocks will be shown. Know where to find each of the terminal blocks, and what circuits are on each block. When a failure occurs, the terminal block is a good place to start looking for the trouble.

Next, inspect the inoperative device for broken or corroded terminals, and check out the switch or breaker on the control panel.

If the trouble can't be located by inspection of the connections, then some testing is required using a meter. While analog meters are effective

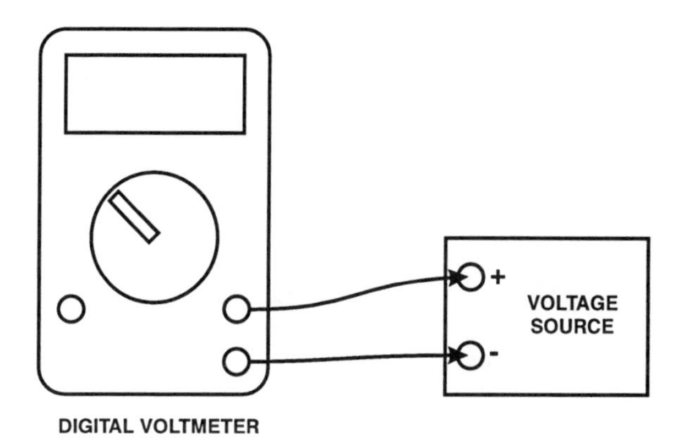

Figure 16.1: Measuring Voltage

tools in the hands of an experienced user, digital meters are economical and more easily used. We assume you have a digital meter.

16.4 Using A Digital Multimeter

Digital multimeters, DMMs, are briefly described in the chapter covering tools. If you are not familiar with the functions of a multimeter, you might reread the section on multimeters.

For most testing purposes, a simple voltmeter is sufficient. Added features that are useful include an Amps function, and a continuity function. A clip–on ammeter is very useful for testing systems that are not completely instrumented. Both AC and DC measurements are required, so make sure the DMM has ranges for both. Another useful feature is the ability to measure diode conductivity.

16.4.1 Measuring Voltage

Voltage is always measured with the two leads of the meter across the voltage source as shown in Figure 16.1. It doesn't really matter which lead of the voltmeter is connected to what terminal of the voltage source. If the red meter lead is connected to the negative terminal with the black lead connected to positive, the only effect will be a minus sign in the meter display. The displayed voltage will have the same magnitude.

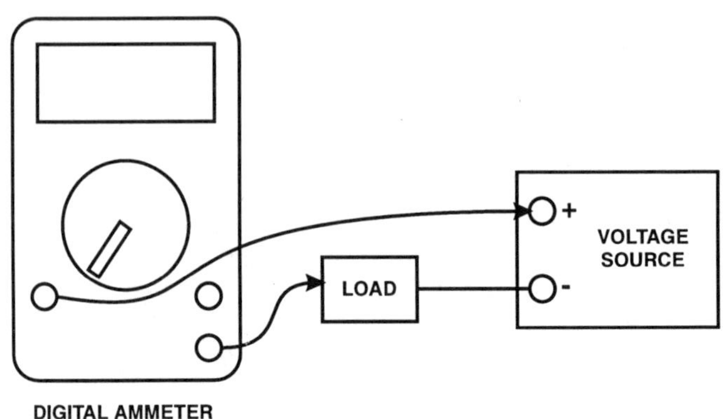

Figure 16.2: Measuring Current

While most meters are protected from accidental overvoltages, it is good practice to always select the proper channel on meters with manual range selection, *before connecting* to the voltage source. Auto ranging meters automatically select the proper range, so overvoltages are not harmful, provided that the meter is on a voltage function.

All too frequently, we use the meter to measure current and forget to reconnect the meter leads for voltage. That always costs us a fuse.

16.4.2 Measuring Amps

Amps can be measured with the ammeter directly, indirectly with a shunt, or with a clip-on meter.

To measure Amps with the multi–meter, the circuit must be broken and the ammeter connected in series with the load as shown in Figure 16.2. Note that the ammeter leads must be connected to the correct terminals for Amps measurement. Most handheld ammeters will only measure up to 10 Amps, so they are rather limited in application.

To measure beyond the 10 Amp limit of the meter, a shunt can be used. The shunt must be placed in series with the load, as was the meter itself. Figure 16.3 shows the proper connections. Note that the meter is set up to measure the voltage drop across the shunt, instead of current. Note also that the leads are connected to the two sensing terminals on the shunt. Measuring the voltage across the main shunt terminals will result in a large error.

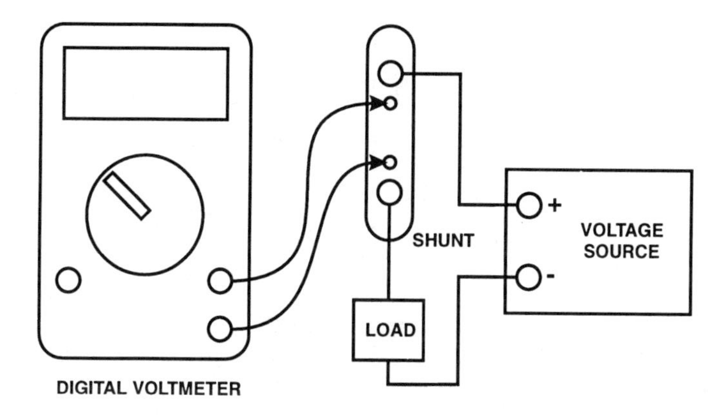

Figure 16.3: Measuring Current with a Shunt

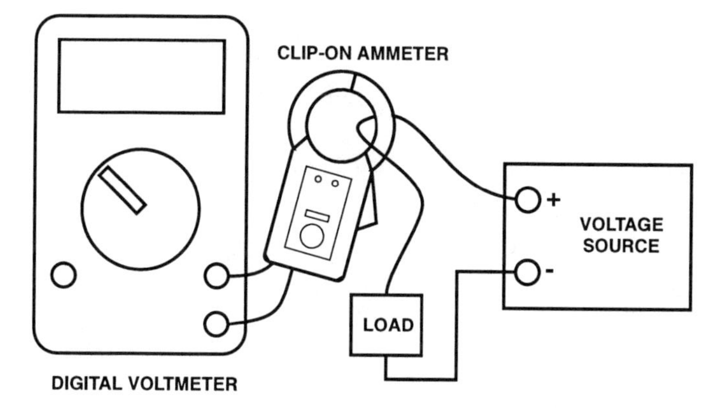

Figure 16.4: Measuring Current with a Clip–On

If a 1 millivolt per Amp shunt is used to measure current, translation of the voltage to the actual current is a 1:1 affair ... every millivolt is 1 Amp. A 1 millivolt shunt is a high voltage shunt. As mentioned, be careful with such a shunt in series with high currents such as starter motor loads.

A clip–on ammeter is most convenient since the circuit need not be broken to insert an ammeter or shunt. Figure 16.4 shows the way to measure current with the clip–on meter. Note that only one wire should be inside the circular jaws at any time. If more than one wire is inside the jaws, the current measured will be the sum or difference of the individual

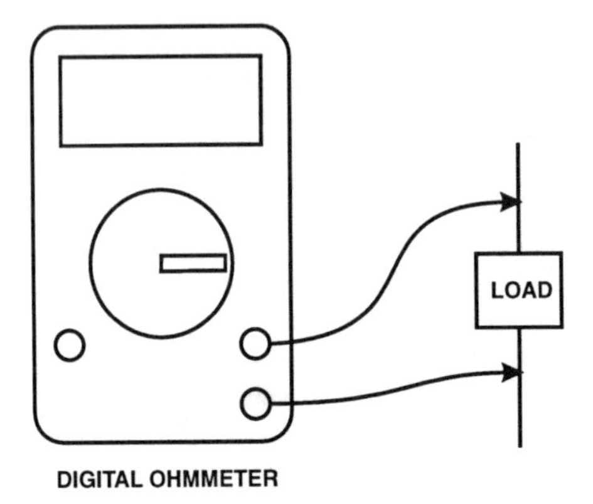

Figure 16.5: Measuring Continuity

currents. If the currents are flowing in the same direction, you'll get the sum of them. For opposite direction currents, the measured value will be the difference.

Be careful to not connect an ammeter across a voltage source. The ammeter is essentially a short circuit, thus it will damage the source, or the ammeter. Modern ammeters are fused, but the fuses are not cheap, and a spare may not always be available. After measuring current, reconnect the meter for voltage measurements.

16.4.3 Measuring Continuity

Figure 16.5 shows how to measure continuity. Continuity measurements are useful to test wires, bulbs, switches, diodes, breakers, transformers, motors, and other devices. Continuity measurements must always be done with no power applied to the device under test. If power is applied, damage to the meter may result. Continuity readings will also be in error, sometimes showing a negative resistance.

Transformers and motors exhibit a very low resistance, often appearing as a short circuit. The field winding on an alternator is also a low resistance, from about 2–8 Ohms. When measuring alternator resistance through the brushes, beware of the high resistance of oil films on the slip rings. The ohmmeter does not generate enough voltage or current to break through the films, so resistance readings are often inaccurate. For best results, the measurement must be done at the slip rings, not through the brushes.

16.4.4 Measuring Diode Continuity

Many DMMs have a selection to measure diode continuity, or conductivity. The DMM applies a voltage to the diode, and measures the voltage drop that develops across the diode. If the meter is connected so that the diode is forward biased (conducting), then the voltage across the diode will be low, say 0.1 to 0.8 Volts. Schottky diodes will measure on the low side, while silicon diodes will measure a higher voltage. The higher the current rating on a diode, the lower it will read when conducting.

If you hook the diode continuity meter in the reverse order, then little or no current will flow, so the meter will indicate an overrange condition. Some low voltage Schottky diodes have a significant reverse current on the diode test, and so they won't read entirely open. Measuring a diode in reverse polarity checks its ability to block current flow ...important if the diode is to be an efficient rectifier. Diodes that conduct in both direction are defective.

More details about checking diodes is given below in the section about testing isolators.

16.5 The Test Path

Earlier, we made the point that electricity must have a closed circuit to flow. Always keep that fact in mind when testing.

Before a load device will operate, it must have voltage applied to it. That voltage comes from a battery, so what you search for is the presence of battery voltage on the circuit under test. Some people suggest that you start testing at the battery and work toward the device under test. Most certainly, you want to verify that voltage is available at the battery to ultimately get to the questionable device. After testing for the presence of battery voltage, it usually pays to then go to the questionable device and test backwards from it to the battery. It is always easier to trace wires from the device back to the battery, than vice versa. This rule might be stated as *verify the source of voltage, and then start at the device.*

A charging device must produce a voltage greater than the battery or it won't charge. In the case of troubleshooting a solar panel, for instance, you would start testing at the panel, seeking a voltage of 13+ Volts. If that test is affirmative, then testing will proceed by working back to the battery along the wires, verifying that the voltage makes it to the battery. If you have small gauge wires, and the solar panel is generating, there can be a significant voltage drop from the panel to the battery. Allow for some drop, but if in doubt, you can always insert your ammeter in series with the panel to measure actual current. In other words, be prepared to measure both voltage and current.

16.6 Testing the Alternator

Unless you have a device that gives an alarm for low voltage, a failed alternator may not be noticed until the batteries are flat. If the batteries are really flat, you won't want to turn the engine off unless you have another means of charging.

Alternator failures are often the result of loose belts or adjustment arms. Even mounting brackets fail, allowing the belt to loosen. High output alternators in particular, require a very tight belt to avoid slippage.

Alternator testing begins by inspecting the wiring that connects to the alternator. Look for corroded or broken terminals. Without an isolator, you should also be able to measure battery voltage on the output terminal of the alternator. If the voltage is either not present, or very high, then the alternator is not connected to the batteries. If you have an isolator, then it may have a loose connection, or it may have failed, opening up the circuit.

A clip–on ammeter is a good tool to have when it is necessary to test the alternator. Even if you have built-in instrumentation, a clip–on ammeter is valuable as a verification that the built–in instrumentation still functions. To use the clip-on meter (see tools section), simply clip the meter around the positive output terminal. Run the engine at a brisk pace (1500 RPM). Turn on a significant current draw, say all the lights in the vessel. The ammeter should show that the alternator is producing current. If not, the alternator, regulator, or connections to the alternator are bad.

With an isolator, the output voltage at an alternator which is not running may read near zero. Since the isolator diodes do have reverse leakage current, it isn't unusual to measure close to the battery voltage at the alternator output post. Refer to the following section for details about testing an isolator.

For alternators with internal regulators, not much testing can be done before removing the alternator from the engine. For externally regulated alternators, the following tests can be made while running. Be careful not to get caught in the belt, and don't cause any short circuits. Shut off the engine before removing the field wire, and only turn it back on when you are prepared to make the tests. To test the alternator itself, it will be necessary to remove the field connection from the regulator, and connect the field wire to either positive or negative, depending on whether the alternator is an N or P type. If the alternator is a P type, then the alternator should charge at its maximum rate whenever the field wire is connected to battery voltage. For a P type alternator, jumper the field momentarily to a battery voltage. If you get no charge, then the alternator is defective. If the alternator does charge when *full fielded*, then the regulator is the likely culprit for lack of charging. An N type alternator is tested by connecting the field to ground ... the alternator should then charge at its maximum rate.

If you can not get any alternator output by full fielding the alternator, then the alternator will have to be removed for bench tests with an ohm-meter. External continuity tests won't show up shorted or open stator windings, but will give you an idea whether the diodes are good. For a schematic of an alternator, refer to *Living on 12 Volts with Ample Power.* It shows the diodes polarities which will be helpful in understanding the following tests.

Start out by measuring the reverse diode current, using the diode function on the meter. Place the positive or red lead on the alternator output post. Connect the negative lead to the case. You should read an open circuit, since the internal diodes are reverse biased. Now switch the polarity of the leads, the red lead going to alternator ground, and the black lead touching the alternator output. The internal diodes are now forward biased, and there are two in series, so expect a reading of about 1 Volt.

The same tests can be done on the tachometer or **R**pm terminal. Put the positive lead of the meter on the R terminal, and the negative on ground. You should read an open circuit. Once again, reverse the leads

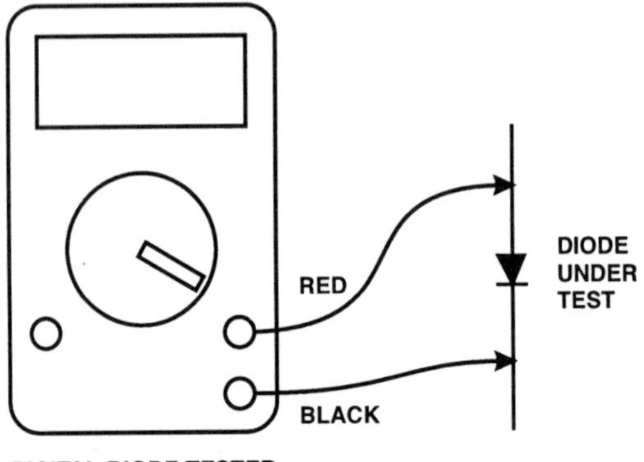

Figure 16.6: Diode Conductivity

and observe the reading. This time, there is only one diode in series, so you should read about half of the value when you tested from the output terminal.

Remember, an alternator can pass the diode continuity test, and still have shorts across a stator winding, or have an open stator winding. Only a full field test of a running alternator can detect these conditions, unless you are willing to break down the alternator and test internal windings.

To measure continuity on the field winding, you may have to separate the cases so that you can bypass the brushes. With the alternator apart, you can place test leads on the slip rings and measure the resistance of the field winding. As mentioned, it should measure about 2–8 Ohms. Puny alternators that produce 30–60 Amps may have slightly higher field resistance.

If a field has less than 2 Ohms, suspect an internal short of the field. A short in the field will require that you replace the rotor.

With the cases apart, individual diodes can be tested for continuity. Be sure to test for reverse blocking capability as well as continuity.

Integrity of the field winding can be tested by applying battery voltage to the alternator field, and measuring the current that flows. Expect 1.5–6 Amps.

16.7 Diode and Isolator Testing

Figure 16.6 shows the connections to be made to test a diode for continuity. With leads connected as shown, the diode test will indicate the forward voltage drop of the diode. With leads connected in the other direction, little, if any diode current will flow. The meter will read open.

For continuity measurements of isolators, you should not have the engine running. You will also need to remove the alternator input wire to the isolator. Isolators should read good when the positive lead of the meter is placed on the alternator terminal, and the negative lead is placed on a battery terminal. Test all legs of the isolator by moving the negative lead to each battery terminal.

After testing for continuity in the forward direction, reverse the meter leads and check for blocking capability. Place the negative lead of the meter on the alternator terminal, and check each battery terminal with the positive lead. The normal reading is an open circuit.

Sometimes diodes will show continuity on the meter, but under the heavy currents of actual operation, the diode won't conduct sufficiently. To test isolator diodes in operation, simply measure the voltage across them. A silicon diode may have as much as 1.2 Volts across it with heavy current flow. Even with low current flow, you may measure 0.7–0.8 Volts. A Schottky diode will exhibit a voltage drop of 0.2–0.8 Volts with low to high current respectively.

16.8 Testing Running Lights

We have found that the cause of most failures of running lights is due to corrosion at the socket. The bulb itself is usually functional if you can measure continuity. Often, you can see that the filament in the bulb is open ... but we've been tricked by appearance, so we usually measure.

Most incandescent light bulbs have a very low resistance when they are cold, virtually a short circuit. If a circuit breaker is *popping* whenever you turn on the running lights, expect to find a short circuit. It is unlikely that the short will be in a bulb, however. Remove the bulbs and retry. If you still blow the breaker, the short is in the wiring.

Intermittent problems are caused by dirty sockets, or weak spring tension on the contacts. Dirty sockets can be cleaned with some fine sandpaper. You can use the striking strip from book matches as a small emery board to clean contacts.

Don't overlook the possibility that voltage is not present at the socket. Wires and breakers upstream can be the culprit.

16.9 Testing Anchor Windlass

The anchor windlass circuit is shown in the chapter of schematics. Note that there is an enabling switch as well as a deck operated pedal switch. These two are wired in series so that both are required to operate the windlass solenoid. The solenoid serves the same function as a starter solenoid, that is, it switches the high currents necessary for the windlass motor. The foot switch only carries enough current to operate the solenoid.

Test for voltage on the foot switch. If present, activate the foot switch and see if the solenoid coil has voltage across it. If it does, and the solenoid doesn't close with a noticeable click, then the solenoid is defective. If the solenoid closes, then there should be voltage at the windlass motor, otherwise, the solenoid contacts are open. If the motor has voltage across it, and still doesn't run, then it must be open circuited.

Switches can also be tested for continuity. Be sure that all power is removed from the switch before testing continuity. Closed contacts should have a resistance of less that 1 Ohm.

16.10 Testing Solar Panels

Solar panels can be tested with the DMM using both the voltage and current functions. Naturally, you need a sunny day to test the panels, and the panels need to be oriented toward the sun. With the voltmeter, test the output voltage of the panel with the panel disconnected from its load. You will be measuring its *open circuit output voltage.* You should get 15 to 25 Volts. If it is less than that, possibly some of the cells are shorted so it doesn't develop enough voltage to charge batteries.

After the open circuit test, and with the panel still disconnected from the batteries, set up the DMM to measure current in the 10 Amp range, and put the meter directly across the solar panel as if you were measuring voltage. The ammeter will act as a short circuit to the panel, which is alright, and it will measure the maximum current flow from the panel.

For rigid panels, expect about 1 Amp for every 200 square inches of panel surface area. Flexible panels will be about half this value, that is 1 Amp for about 400 square inches of surface.

If the solar panel tests good, and you still aren't charging the batteries, suspect the wiring between the panel and the batteries. As usual, the connectors are the most likely suspect, with terminals next in order of failures.

16.11 Testing Wind/Water Generators

Wind and water generators are most often DC motors which are driven, and therefore become generators. While they may be tested without removing them from their normal mounts, it is usually easier to remove the generator and test it with an electric drill. Simply chuck the generator shaft into the drill motor and turn on the juice. With the generator leads not connected, you should measure a relatively high voltage, perhaps as much as 100 Volts at the output.

We don't suggest measuring short circuit current of the generator using the standard DMM ammeter, since the maximum DMM scale is typically only 10 Amps, and the average wind/water generator can produce more than this if it is working properly. A quick test of output can be done by shorting the output leads and then trying to run the drill motor. Unless you have a large drill motor, it won't be able to run the generator more than a few RPM. Big generators can stall a small drill motor. Don't run the generator very long with the leads shorted ... permanent damage can be done if run too long with leads shorted.

16.12 Testing Refrigeration Controls

Operation of the typical refrigeration system is quite simple. A thermostat closes to activate the compressor motor, and opens when the evaporator temperature falls low enough to open. The motor can be tested by shorting the thermostat contacts with an alligator lead. The motor should run with the thermostat shorted. The thermostat contacts can be

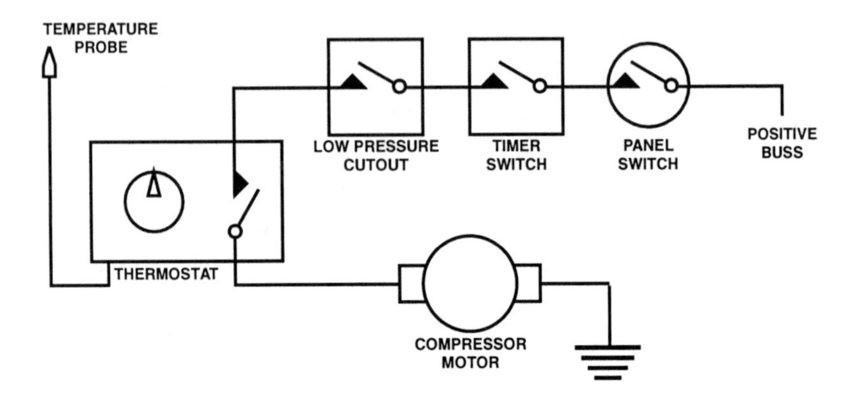

Figure 16.7: Holdover Refrigeration

measured with the Ohms function in the DMM. Disconnect the leads on the thermostat before connecting the ohmmeter across the thermostat. At room temperature, the thermostat should read a low resistance, less than 1 Ohm.

The thermostat has to be cold to open. If you have access to a functional freezer, you can test the thermostat in it.

If the thermostat tested closed at room temperature, re–connect it and turn on the system. The motor should run with the thermostat closed. If it doesn't you might try the direct application of voltage to the motor, using alligator leads.

If the motor runs, but the box doesn't get cold, then the system is probably low on refrigerant.

If the motor continues to run after the box is cold, measure the voltage drop across the thermostat contacts. You should read battery voltage if the thermostat has opened. Disconnect one lead from the thermostat and check to see if the motor stops. A shorted wire to the motor can cause it to run continuously, and it may appear that the thermostat is not opening.

Figure 16.7 shows typical controls for a holdover plate system. The unit is usually started by a switch, which may be a timer switch. Wired in series with the control switch is a low pressure switch, and often a high pressure switch. The low pressure switch is normally closed, and opens whenever the holdover plate has reached a low temperature. The high pressure switch, if present, is a safety switch that opens if pressure on the condenser side rises to dangerous levels. The control switch with the other pressure switches either control a clutch on the main engine, or an electric motor. Electric motors may be wired through a relay if motor current is greater than the capacity of the switches. A relay, if present, will activate when all switches are closed. With the relay activated, the electric motor should operate. See the section on relay testing for more details about relays.

16.13 Testing Battery Chargers

As usual, check the connections on the battery charger for signs of mechanical failures.

Most battery chargers must be connected to the battery before their output can be tested with a voltmeter. The typical battery charger outputs pulses of current which do not register correctly on the meter unless a battery is in the circuit to average the pulses. For bench testing, a large capacitor of several thousand micro-Farads can be substituted for the battery. To more closely simulate a battery, a load of about an Amp should be placed in parallel with the capacitor.

With the charger connected to a battery, a quick test of output is possible by measuring the battery voltage with the charger off, and then again after turning the charger on. Except for a huge battery that is deeply discharged, you will see a higher battery voltage with the charger on. This test doesn't tell you much about the actual output of the charger ... a test of current output is required for that. To test the output current, you will either have to interrupt the circuit by disconnecting the negative lead, or you will have to use a clip–on ammeter. Obviously, the clip–on meter is faster.

If you have an AC ammeter in the circuit, you can often get some indication of charger output by observing AC current. Turn the charger on and off and observe the difference in AC current. As a rule of thumb, if you multiply the AC current by 10, you will have an idea of charger output current.

By all means, don't leave batteries connected to a charger unless you have tested the maximum voltage that the charger goes to when the batteries are fully charged. As mentioned, ferroresonant chargers are notorious for overcharging if left on long enough. We suggest throwing such chargers overboard and replacing it with a good charger, but if economic constraints force a less radical solution, be sure to measure the maximum output voltage before trusting the care of expensive batteries to the ferroresonant charger. Maximum output voltage can only be measured when the batteries have taken a full charge. Because ferroresonant chargers output very little current at the full mark, there must be no loads attached to any of the batteries during testing of maximum voltage.

To measure maximum voltage, leave the charger attached to the batteries for several days with no load connected. Now, simply measure the battery voltage at each battery. If any of them are above 13.8 Volts, throw the charger away, or use it sparingly, only when you are aboard for the weekend.

Ferroresonant chargers are also very sensitive to AC line frequency, and voltage, so you really need to make the maximum voltage test at high line voltage, say 125 VAC. The frequency of power from the public utility is precisely controlled, but if you are running a charger from an auxiliary generator, be sure it is producing 60 Hz.

As often as not, ferroresonant chargers don't produce a high enough voltage. This means you will not charge batteries fully. If the batteries are consistently charged only to about 13.5 Volts, they won't be fully charged. You will need a means to periodically bring the voltage to about 14.4 Volts, and maintain that voltage until battery current has fallen to about 2% of its Ah rating.

SCR phase controlled chargers are adjustable. Like the ferroresonant charger, they can only be tested when connected to a battery, and can only be adjusted after the battery is fully charged. For maximum battery life, adjust the output to about 13.65 Volts. That won't fully charge the batteries, so have an alternate way of restoring full charge. Periodically, the charger can be adjusted to 14.4 Volts and then down again after an absorption cycle. A 3–Step Regulator on the alternator can produce a full charge whenever the engine is run.

Not all battery chargers need to tested with batteries attached. Chargers that produce pure DC can be tested and adjusted without batteries. This can be useful when you must adjust a unit, and time doesn't permit waiting for the batteries to get fully charged. Simple disconnect the negative lead on the charger, and adjust the output voltage as desired.

Reconnect the negative lead ... the batteries will eventually charge to the adjusted voltage.

Battery chargers can often lead to electrolysis problems if they develop internal shorts. A charger should have no continuity between the case, and any of the output leads. Before testing for this, disconnect the charger from both AC and DC circuits.

Not all shorts or partial shorts will show up with external tests. We have seen chargers develop internal leaks to the case that didn't show up from simple tests at the output leads. Chargers with cases that are riveted together prevent internal tests unless you are willing to void the warranty. If zincs are being consumed at a high rate, suspect the charger. Disconnect all wires to it for a few months to see if it is the cause.

16.14 Testing Relays

Relays consist of an electromagnet coil, and one more set of contacts. Relays are made with DC coils and AC coils. A DC relay operated on AC will probably chatter as it opens and closes 120 times per second. An AC relay operated on DC may not actuate at all, or it may actuate momentarily and then open again. In either case, overheating is likely, with failure to follow.

Relays are also made to operate with various voltages. Before applying a test voltage, be sure that it is appropriate for the relay under test. Often, the schematic for a relay is shown on the relay package, and generally, the coil voltage is given as well. Schematics are always drawn with no voltage applied to the coil, so contacts shown closed should be closed with no connections to the relay. Contacts can be measured using the ohmmeter function of the DMM. Resistance should be less than an Ohm for closed contacts, and an open circuit for open contacts. Measure all contacts before coil actuation. Don't trust normally open contacts to be open. Contacts can be welded closed with sufficient current, which happens with short circuits or excessive motor loads.

After testing all the contacts with no coil power, apply the proper voltage to the relay coil. As mentioned in the section on tools, a few alligator clips are handy things to have when testing electrical parts. You should hear a click as the relay wiper transfers to the normally open contacts. Measure all the contacts again. Contacts previously open should be closed, while those closed should now be open.

It is rare that shorts develop between adjacent poles and contacts, but it doesn't hurt to test contacts of one pole for shorts to contacts of another pole.

16.15 Testing Autopilots

We've known single handed sailors that somehow manage without an autopilot or a windvane. One friend even goes without a motor, using a large sculling oar instead. We are of the opinion that a boat without an autopilot is not fully functional. This is especially true of large boats that are shorthanded.

Our autopilot has about 8000 miles of steering to its credit, but not without a few repairs. The first failure occurred after two days of downwind sailing to Mexico. The shaft on the rudder feedback potentiometer froze up, causing the potentiometer to strip loose from its printed circuit board mounting. We managed to solder on some wires while underway and get it operating.

A replacement potentiometer was shipped into Mexico to us, but it also suffered the same failure after an overnight trip into the Sea of Cortez. This time we looked at the mounting assembly carefully and discovered that the shaft was misaligned, and there was nothing in the coupling that could flex. Rather than order another potentiometer, we asked for a schematic from the manufacturer. We rooted around our assortment of electronic parts and came up with a potentiometer that had a plastic shaft. It wasn't the same value, so a few other resistors had to be changed before it was functional. The plastic shaft held up and might still be working today, but we replaced the whole rudder feedback assembly with our own design once back in the U.S.

While in Mexico, the compass location was changed to allow the watch better access. In the process, the compass card raised up off its pivot. It still worked, as long as we weren't sailing at 180°. At that point of sail, it would bind slightly on the case, and we'd end up sailing S curves. By taking the compass apart, the problem became obvious, but the cure wasn't. The card was enclosed inside a plastic sphere which was glued to its mount. At sea, we drilled a couple of toothpick size holes in the sphere and was then able to manipulate the card back onto its pivot.

The autopilot held up from Mexico to Hawaii, and most of the way back to Seattle. The compass card jumped its pivot when we got pooped a few days from Seattle, but it wasn't obvious until one night we decided to sail south to miss a storm center bearing down on our course. We got the card seated, but spilled a little compass oil in the process. Teak oil was all we had on board that was close to the same viscosity, so that's what we used.

On our first trip to the San Juan Islands, the autopilot quit again. This time it was the relay that reverses current to the motor. Yet later, the plastic sphere in the compass came unglued. That could have been from stress on the sphere whenever we drilled it, or from prying with toothpicks to seat the card. It might also have been from the teak oil that we used to top off the container when we reseated the card during the trip between Hawaii and Seattle. We wiped the sphere, and mount free of oil by cleaning with ordinary rubbing alcohol, and stuck the two together with epoxy. We reused the teak oil mixture.

Even when the autopilot worked, it didn't really. For brisk downwind weather with following seas, we either had to steer, or reduce sail. The autopilot is once again broken. We haven't bothered to troubleshoot it ...it will be replaced with our own design before any long cruises.

As expected, the failures that we encountered with the autopilot involved mechanical moving parts. First, the feedback potentiometer, then the compass card, next a relay. The motor, which drives a hydraulic pump, always worked, and still does, but a motor can be a weak link.

When the rudder feedback potentiometer failed, the autopilot would oversteer, and then oversteer again when its error became apparent. This meant giant S turns at any point of sail.

The sticky compass card caused S turns, but only in one direction. The relay failure only allowed motor rotation in one direction, which makes for pretty boring cruising.

For a completely inoperative unit, start by verifying that power is getting to the controller. If power is present, and the unit has no reponse, check out power at the motor. If it has power and doesn't turn, the motor is bad, or it may have a thermal overload breaker that has opened. If power doesn't get to the motor, look for a reversing relay. Newer autopilots may not use a relay, but instead use a *transistor bridge* to reverse motor current. As always, keep your eyes open for failed connections.

16.16 Tech Tips

16.16.1 General Information

The following subjects are not in any special order. Each subject is a method that allows quick testing for functionality, or wiring.

16.16.2 Determining Which Battery

If you sometimes go onto strange boats and wish to know or verify which physical battery bank is connected to selector switch position 1 or 2, simply grab the digital voltmeter. Turn off all the loads and measure the voltage on each bank. Now, turn the battery selector switch to position one and turn on some loads, say running lights. The selected bank should show a steady decline in voltage. The bank that shows a decline in voltage is bank #1. If both banks decline, they are somehow connected together. Is the selector switch in both? Did an amateur electrician run a wire from the alternator to both batteries?

16.16.3 Is the Anchor Light On?

Another useful application of Amps measurement is to determine if the anchor light is on. Often, we have left the boat swinging on the hook in the afternoon, with time of return expected after dark. With accurate measurement of Amps, it is simple to know that the anchor light is actually burning, even though the light is not visible. Even without knowing exactly the Amps that the bulb takes, turning the anchor light on and off will show a difference of current flow if the light is working.

16.16.4 Voltage Under Load

As indicated above, the voltage of a loaded battery declines from its rested voltage when a load is turned on. When you turn on a switch for a pump, and the pump doesn't run, is it drawing current? If you have an Amps meter, simply observe the meter when the switch is turned on and off. Without an Amps meter, observe the voltage of the battery. If it falls whenever the switch is on, then the pump is drawing current. If the pump isn't running, then the motor must be locked. If circuit breakers are sized correctly, a locked motor will pop the breaker.

16.16.5 Finding the Neutral Wire

Caution, this test presents quite an opportunity to get electrocuted. Be sure that you understand the test procedure, and do not make contact with any of the circuits being measured.

My policy is to never trust a marina slip to have the AC plug wired correctly. Often, the black and white wires will be switched. If you have a polarity light on board, a switched black and white wire shows up immediately. But, can you detect a switched neutral (white) and ground (green)? Probably not. With a voltmeter, it is easy to check for a crossed white and green wire. You'll need a scrap piece of wire, or else you'll have to dip one of the voltmeter probes into the water. Clip one of the voltmeter leads to the scrap wire, and dangle an exposed piece of the wire into the water. On the 300 Volt AC scale, measure the voltages from each of the wires in the shore power outlet. The one with the least voltage is the safety ground (green wire). The next slightly higher voltage is the neutral (white) wire. Finally, you should find a wire that measures about 115 VAC with respect to the water.

16.16.6 When the Engine Won't Start

One night on the way back to Seattle from Hawaii we were pooped by a giant wave and knocked down. The ignition switch on the helm got doused. A couple of days later, in the middle of the night, the engine started by itself! We cut the wire connected to the starter solenoid and jury–rigged a toggle switch in line.

Normally, the problem is failure of the engine to start. Most engine starting systems include a starter solenoid that is activated by the ignition switch. The solenoid switches the high current necessary for the starter motor.

When the engine won't crank, the failures can be in several places. Does the ignition switch activate the starter solenoid? Does the starter solenoid activate the starter motor? Is the engine grounded to the battery negative? These questions are easily answered with a digital voltmeter. Begin by connecting the negative voltmeter lead to battery ground. Place the positive lead on the engine block. At most, you should only read a few millivolts difference. Turn on the ignition switch to the start position, with the leads still connected. Again, there should only be a few millivolts if the engine ground is intact. Move the positive lead of the meter to the big positive lug on the starter motor. Again activate the starter switch to engage the starter. If the solenoid is working correctly, you should measure battery voltage at the starter. If you do measure full battery voltage, then the starter motor is defective. If you only measure partial battery voltage, look for a poor connection in either the positive or negative connections to the engine. If you measure no voltage at the main starter terminal, the ignition switch or the starter solenoid is not functioning properly. Continue testing.

The starter solenoid will have a large terminal that has battery voltage on it from the selector switch. The other large terminal will go to the starter motor. A smaller terminal will be connected to the ignition switch. This small terminal should have battery voltage applied to it whenever the ignition switch is in the start position. If not, the ignition switch is defective, or does not have battery voltage connected to itself. If the starter solenoid has battery voltage on it whenever the starter switch is in the starting position, and the solenoid doesn't switch the battery voltage at its large terminal over to the starter, then the solenoid is defective. Be sure that battery voltage does come into the solenoid on one of its big terminals. If not, you'll have to trace the wiring back toward the battery selector switch.

Bibliography

David Smead and Ruth Ishihara, "Living on 12 Volts with Ample Power", First Edition, Rides Publishing Company, 2442 NW Market Street, #43, Seattle, Washington, 98107.

Issac Asimov, "Asimov's Biographical Encyclopedia of Science and Technology: the Lives and Achievements of 1510 Great Scientists from Ancient Times to the Present Chronologically Arranged", Second Revised Edition, Doubleday and Company, Inc.: Garden City, New York, 1982.

Issac Asimov, "Chronology of Science & Discovery". How science has shaped the world and how the world has affected science from 4,000,000 B.C. to the present, First Edition, Harper & Row, Publishers, Inc., 10 East 53rd Street, New York, NY 10022.

M. Barak, "Electrochemical Power Sources: Primary and Secondary Batteries", First Edition, Institute of Electrical Engineers (IEE) Series 1, The Institute of Electrical Engineers: London and New York, Peter Peregrinus Ltd., Stevenage, UK, and New York, 1980.

A. E. Fitzgerald, Charles Kingsley, Jr., Alexander Kusko, "Electric Machinery: The Processes, Devices, and Systems of ElectroMechanical Energy Conservation", Third Edition, McGraw-Hill, Inc.: New York, 1971.

G. Smith, "Storage Batteries: Including Operation, Charging, Maintenance and Repair", Second Edition, Pitman Publishing: Great Britain, 1971.

William M. Flanagan, "Handbook of Transformer Applications", McGraw-Hill, Inc.: New York, 1986.

E. M. Pugh, E. W. Pugh, "Principles of Electricity and Magnetism", Second Printing, Addison-Wesley Publishing Company, Inc.: Reading, Massachusetts, 1962.

Conrad Miller and Elbert S. Maloney, "Your Boat's Electrical System; Including A Manual of Marine Electrical Work", Second Edition, Hearst Books: New York, 1981.

Electro-Craft Corporation, "DC Motors, Speed Controls, Servo Systems," Fifth Edition, Electro-Craft Corporation: Minnesota, 1980.

General Electric, "The Sealed Lead Battery Handbook", Publication BBD-OEM-237, General Electric Company: Gainesville, Florida, 1979.

American Boat and Yacht Council, "Standards and Recommended Practices for Small Craft": Amityville, New York, 1987.

David Linden, "Handbook of Batteries & Fuel Cells", McGraw-Hill Book Company: New York, 1984.

Underwriters Laboratories, "UL 1112: Marine Electric Motors and Generators (Cranking, Outdrive Tilt, Trim Tab, Generators, Alternators)": Northbrook, Illinois, 1983.

Underwriters Laboratories, "UL 1236: Battery Chargers": Northbrook, Illinois, 1986.

Ralph G. Hudson, S.B., "The Engineers' Manual", Second Edition, John Wiley & Sons, Inc.: New York, 1939.

Index

Living on 12 Volts with Ample Power

David Smead and Ruth Ishihara are the authors of *Living on 12 Volts with Ample Power*. Published in 1988, the book quickly became the definitive reference for battery energy systems. Seven years, and six printings later, *Living on 12 Volts with Ample Power* is still the definitive and authoritative reference. It provides the information you need to own and operate a small energy system aboard a boat, RV or in a remote home.

Living on 12 Volts with Ample Power provides detailed information about liquid electrolyte batteries, immobilized electrolyte batteries, DC alternators, battery chargers, solar panels, wind and tow generators, the AC system, compressor refrigeration, non–compressor refrigeration, electrolysis, and AC safety. It also covers electrical symbols and circuits with historical information about some of the electrical pioners.

Living on 12 Volts with Ample Power .... $25.00.
Wiring 12 Volts for Ample Power .... $20.00.
Both Books .... $40.00.

To order, send check or money order to:

Rides Publishing Company
2442 NW Market Street #43W
Seattle, WA 98107.